Current Topics in Microbiology and Immunology

256

Springer
Berlin
Heidelberg
New York
Barcelona
Hong Kong
London
Milan
Paris
Singapore
Tokyo

Hantaviruses

Edited by C.S. Schmaljohn
and S.T. Nichol

With 24 Figures and 14 Tables

Springer

Connie S. Schmaljohn, Ph.D.
Molecular Virology, Virology Division
USAMRIID
1425 Porter Street
Ft. Detrick
Frederick, MD 21702-5011
USA
E-mail: connie.schmaljohn@det.amedd.army.mil

Stuart T. Nichol, Ph.D.
Molecular Biology Section
Special Pathogens Branch
Division of Viral and Rickettsial Disease
Centers for Disease Control and Prevention (CDC)
Atlanta, GA 30333
USA
E-mail: stn1@cdc.gov

Cover Illustration: Background: Immunofluorescent antibody staining of Seoul virus-infected Vero E6 cells with polyclonal sera to Seoul virus. Foreground: Schematic representation of a hantavirus particle showing the three genome segments complexed with the nucleocapsid and polymerase proteins and surrounded by a bilaminar envelope with two protruding viral glycoproteins.

ISSN 0070-217X
ISBN 3-540-41045-7 Springer-Verlag Berlin Heidelberg New York

Springer-Verlag Berlin Heidelberg New York
a member of BertelsmannSpringer Science + Business Media GmbH

Library of Congress Catalog Card Number 15-12910

Cover Design: *design & production GmbH*, Heidelberg
Typesetting: Scientific Publishing Services (P) Ltd, Madras
Printed on acid-free paper SPIN: 10718346 27/3020 5 4 3 2 1 0

Preface

Hantaviruses hit the headlines in 1993 when a large outbreak of hantavirus pulmonary syndrome (HPS), a previously unknown human disease with high mortality, erupted in the southwestern United States, and was quickly shown to be caused by a newly recognized virus, later named Sin Nombre virus, maintained in deer mice populations. Since this time there has been an exponential growth in studies and knowledge of these fascinating viruses, and many of these enlightening studies are highlighted in this volume. However, the roots of the field can be traced back to the early 1950s when United Nations troops were engaged in the Korean conflict. Over 3000 cases of an acute febrile illness were seen among the soldiers, about a third of which had hemorrhagic symptoms, and 5–10% died. The disease was initially called Korean hemorrhagic fever, but is now known as hemorrhagic fever with renal syndrome (HFRS). It took almost 25 years to identify the virus and the rodent reservoir, the field mouse, *Apodemus agrarius*. The virus was named Hantaan virus after the Hantaan river, which runs along the 38th parallel, and is close to the region where *Apodemus* first shown to harbor the virus were trapped. Hantaan virus became the prototype of the genus *Hantavirus* in the family *Bunyaviridae*, and included the only known members of the family that had no arthropod vector. Shortly after the isolation of Hantaan virus, other HFRS-associated viruses such as Seoul virus in rats and Puumala virus in bank voles were identified. In addition, a hantavirus that was not known to be pathogenic to humans, Prospect Hill virus, was discovered and isolated from infected meadow voles in the United States. These discoveries set the stage for the finding, more than ten years later, of the first pathogenic hantaviruses known to exist in the Americas. During the past seven years, HPS has been shown to occur throughout the Americas from Canada to Patagonia, and at least 10 different HPS-associated hantaviruses, each associated with a distinct rodent species, have been discovered. With the study of the many HFRS- and HPS-associated hantaviruses, an increasingly complex picture has emerged with a spectrum of additional hantaviruses of unknown human pathogenicity detected in a variety of other rodent species.

This volume opens with a chapter by Karl Johnson describing the fascinating disease detective work done in the early years of discovery of these viruses. The next two chapters by Colleen Jonsson and Connie Schmaljohn, and Christina Spiropoulou, present a comprehensive overview of the recent advances in understanding hantavirus replication and maturation. Chapters 4 and 5 by Alexander Plyusnin and Sergey Morzunov, and Brian Hjelle and Terry Yates, describe the intricate long-term relationship that hantaviruses have maintained with their specific rodent reservoirs. Unlike the acute infection seen in humans, hantavirus infection of its primary rodent reservoir results in persistent infection with prolonged virus shedding in excreta. Also, phylogenetic data suggest these viruses have been associated with their rodent hosts for millions of years. Like other zoonotic diseases, studying the process of virus maintenance and transmission within rodent communities is critical to understanding hantavirus disease ecology and developing the ability to predict locations and periods of high risk of human exposure. Chapter 6 by Eric Mackow and Irena Gavrilovskaya describes exciting developments in attempts to identify host cell receptor usage and correlate these with viral pathogenicity. In Chapters 7 and 8, Delia Enria, Ana Briggiler, Noemi Pini, and Silvana Levis, and Mats Linderholm and Fredrik Elgh, provide up to date clinical descriptions of HFRS and HPS. The rapid progression and frequent severity of these diseases, make early detection and careful management of patients critical. The detailed descriptions of symptoms and disease parameters should be useful to clinicians that may encounter hantavirus-infected patients. Similarly, in chapter 9, Olli Vapahlahti, Åke Lundkvist, and Antti Vaheri review important recent information correlating major histocompatibility alleles of HFRS and HPS patients, virus load and virus-specific antibody levels with severity of disease. Modulating this complex interaction of virus and host immune response is the focus of the final chapter by Jay Hooper and Dexin Li, which describes recent innovations in the development of vaccines for HPS and HFRS.

We have come a long way since the early days of the hantavirus field in the 1950s. The pace of discovery has accelerated enormously in the last 5–10 years. The application of polymerase chain reaction technology has allowed the rapid identification of an increasingly numerous and complex rogues gallery of disease-associated hantaviruses. The realization of the global nature of hantavirus-associated diseases, and the increasing application of technologically advanced approaches to hantavirus study will further boost the rate of discovery and depth of knowledge of this important group of viruses and their associated diseses.

October 2000

Connie S. Schmaljohn
Stuart T. Nichol

List of Contents

List of Contributors

(Their addresses can be found at the beginning of their respective chapters.)

Hantaviruses: History and Overview

K.M. Johnson

1 Introduction

As early as 1913, Russian clinical records from far eastern Siberia describe the human viral disease now known as hemorrhagic fever with renal syndrome (HFRS) (Casals et al. 1970). Dr. Ho Wang Lee (1982) found a Chinese medical account of a similar disease dating to about A.D. 960. This is not surprising in view of its unique renal complication and the fact that the etiologic agent of HFRS proved to be a zoonotic virus that causes chronic infection with urinary excretion in the murid rodent, *Apodemus agrarius*. This mouse is the most common wild rodent over much of northern and eastern Asia and often invades cultivated fields, gardens, haystacks, and sometimes human dwellings. Human conflict during the lethal 20th century also played a prominent role in elucidation of the etiology, clinical evolution, epidemiology, and ecology of HFRS. "Field nephritis", which occurred in both allied and German troops in Flanders during World War I, may well have been caused by a hantavirus now known to occur in western Europe and Scandinavia (Bradford 1916; Arnold 1944). Japanese military physicians encountered the disease in the mid 1930s after invading Manchuria (Kitano 1944), Finnish and German soldiers were affected during World War II (Stuhlfauth 1943; Hortling 1944), and UN troops first encountered the original hantavirus during the Korean Conflict in 1951 (Smadel 1953).

Departments of Biology and Medicine, University of New Mexico, 10 Calle Final, Placitas, NM 87043, USA

The recent history of hantaviruses and the human diseases they cause is a tapestry being woven at a furious pace. No zoonotic viruses of wild animals causing human disease are as widely distributed on earth as are members of this genus, and research on these agents is pursued by more investigators in more countries than for any other such virus group. Thus, the history of hantaviruses could already fill a generous monograph. The following summary chronicles the most important phases of that history.

2 Early Disease Descriptions (1930s–1960s)

Japanese physicians first encountered HFRS in Manchuria in 1934. Focal small outbreaks occurred among troops during field maneuvers, and wild rodents were suspected as possible reservoirs. By about 1940 detailed clinico-pathological descriptions of the syndrome had been completed. The incubation period of the disease was estimated as 2–3 weeks. Hemoconcentration and a peculiar "leukemoid" white cell response were documented, as were various hemorrhagic phenomena, a hypotensive crisis at the time of febrile resolution, and a subsequent phase of renal insufficiency. Mortality was often more than 10% (Ishii 1942). A similar, though clinically milder, rural disease was recognized during the 1930s in both Scandinavia and in central Russia (Myhrman 1951; Smorodintsev et al. 1959). Scandinavian authors named it nephropathia epidemica (NE).

Japanese scientists utilized human subjects in an effort to isolate a causative agent for HFRS. Tissues from wild mice (*A. agrarius*) and from mites (*Laelaps jettmari*) induced the disease in volunteers, but no other animal could be established as an experimental host for infection (Kitano 1944). Subsequent events showed that this work was half correct (rodent reservoir) and half misleading (no arthropod is involved in virus transmission). Soviet scientists were also busy during this period. Annual outbreaks of HFRS were seen in the Amur River Valley of Siberia beginning in 1932. In studies utilizing human subjects, Smorodintsev (1959) concluded that three species of mites contained a filterable virus that reproduced the disease. Again, no laboratory host was found in which serial replication of the presumptive agent could be demonstrated. Meantime, Soviet epidemiological studies incriminated field mice ever more clearly. In one 1939 study, workers in two camps 4km apart were engaged in earth moving labor about 3km away from the camps. Thirty-one workers in one camp, and only those living closest to a refuse dump, acquired infection, while only one of those in the other camp sickened. Rodent contact was documented to be highest in and near the "infected" tents of the camp where disease occurred (Smorodintsev 1959). Yankovski and Povalishina (1967) concluded that the incubation period of disease could be as long as 6 weeks, that disease in eastern Siberia had spring and fall peaks among persons mainly exposed in forested areas and along river courses, and that late fall and winter were times when the European form of disease was greatest.

There were 3- to 5-year cycles of virus activity, and these were correlated with observed population cycles of field rodents, principally the red-backed vole, *Clethrionomys glareolus*. Furthermore, it was postulated that human infection was most often acquired by inhalation of dust contaminated by rodent urine and feces, not by the bite of any arthropod. The most dramatic experience favoring this hypothesis was reported by KULAGIN et al. (1962) and has been described in full detail in a monograph by TRENCSENI and KELITI (1971), which reviews the clinical and epidemiological features of HFRS in Eastern Europe and the Soviet Union. In September 1961, many rodents, including *Clethrionomys glareolus* and *C. rutilus*, were sent from Kirov county, Russia, to a laboratory in Moscow as part of a study of tick-borne encephalitis. After two weeks in quarantine, these mice were placed in two rooms of one of two sections separated by a central hall. The first human case of HFRS occurred on 18 October, and by 1 November thirty cases had been recorded. At that time, 20 of 23 workers in the section where the mice were housed had fallen ill, whereas only 10 of 40 from the other section were affected by that time. Five employees had experienced the disease in prior years, and none of these became ill during this epidemic. Eventually, 52 of the 58 susceptible workers were clinically infected. Employees working in other buildings also became ill, with the frequency of such illness being a function of the number of visits. Finally, disease began to occur in casual visitors to the laboratory (44 of 94 persons exposed). In total, 113 of 186 workers and visitors who were susceptible to infection acquired the disease. Fortunately there were no fatalities. All mice were sacrificed on 2 November, and the last case was noted on 29 November. Thus, at least for the European form of HFRS, aerosol transmission from presumably chronically infected mice, incubation periods of up to 32 days, and solid immunity to reinfection among personnel who had experienced similar illness in the past, were all convincingly documented.

American medicine met HFRS head-on in 1951 in Korea. By 1954, about 3,000 soldiers had been clinically diagnosed with this disease. The Japanese literature was rapidly translated during the early months of this interval and the term epidemic hemorrhagic fever, favored in that country, was adopted. A special field hospital and associated research laboratory were established to which suspect patients were evacuated by helicopter. In this regard, American physicians were unknowingly following advice of Professor SMORODINTSEV (1959) who wrote, "Such patients tolerate shaking badly and their transport should therefore be gentle". Although no etiologic agent was identified in these studies, the intensive clinical, physiological, and pathological investigations carried out provided a description of this disease more detailed than any single study of a new syndrome, ever. This work is contained in 12 articles comprising 91 pages published as a symposium in the American Journal of Medicine (EARLE 1954). Mortality was about 7%. Epidemiological studies essentially confirmed earlier work on rodents by Russian and Japanese authors, and clearly showed that both spring and fall peaks of disease occurred in the region near the 38th parallel where illness was concentrated.

A detailed study was made by TRAUB et al. (1954) of ectoparasites on four species of rodent found in the region. *Apodemus agrarius* was the most common of these. Ticks and fleas were seldom encountered and were discarded as likely virus vector candidates. *Laelaps jettmari* was the leading laelaptid mite encountered and was most numerous on *Apodemus*. There was no seasonal pattern of quantitative parasitism, so it was (correctly) discarded as a candidate. Trombiculid mites were the most common ectoparasites of all rodents, and they also exhibited a seasonal pattern of abundance that mirrored that of the human disease. But these mites were never found to be engorged with blood or lymph, and none of the human patients had signs of arthropod bites, facts that caused the authors to doubt that these "chiggers" played any role in virus transmission to man.

LAHDEVIRTA (1971) reported and reviewed studies of HFRS in Finland and Scandinavia since 1934. In Finland, most disease occurred during the months from October to March with a clear association between human disease and high rodent populations. Several species of mice, including red-backed voles, were noted to invade barns, other outbuildings, and human dwellings during winter months; the colder the weather the greater the invasions. As observed in European Russia, the disease was generally not fatal. Clinical, physiological, and pathological studies done in Finland were detailed and stand as a solid counterpoint to those of the more severe form of the syndrome done in Korea.

The 1950s saw recognition of HFRS in other countries of Eastern Europe including Yugoslavia, Bulgaria, Czechoslovakia, and Hungary (GAJDUSEK 1962). The Hungarian work of more than a decade was assembled in a monograph by TRENCSENI and KELITI (1971). This work contains the first reference this author can find to the use of dialysis for treatment of severe renal insufficiency. Mortality in Hungary ranged to 10%, suggesting that the Asian, as opposed, or in addition, to the European form of infection was present. As will be cited below, the authors correctly identified the primary rodent hosts of the Asian, Western European and Eastern European forms of HFRS in a table listing "dominant" rodent species for these regions: *Apodemus agrarius*, *Clethrionomys glareolus*, and *Apodemus flavicollis*, respectively.

3 The First Hantavirus

Investigators at the National Institute of Allergy and Infectious Diseases laboratory in the Panama Canal Zone isolated a virus (Machupo) causing acute hemorrhagic fever in eastern Bolivia in 1963 (JOHNSON et al. 1965a,b). Chronic infection of the rodent *Calomys callosus* with continuous urinary shedding of virus was documented. An eradication campaign in the town of San Joaquin, where more than 300 cases of illness had occurred during an interval of 18 months, resulted in complete control of the disease. Although extensive search for infection of *Calomys* ectoparasites was done, a virus could not be recovered (KUNS 1965). With that background, I made visits to the Soviet Union in 1966 and 1969 as part of a US delegation to learn more about several Soviet hemorrhagic fever syndromes.

Although the Bolivian disease did not evoke the renal pathology observed in Eurasia, and leukopenia instead of a leukemoid reaction characterized the South American disease, the epidemiology of HFRS, especially the apparent rodent reservoir with plausible chronic excretion of virus, resonated strongly in my mind. The long incubation period, the powerful lymphocytic leukocyte response, and the inability of anyone to recover an agent from clinical samples should have cautioned me that it was unlikely that HFRS would prove to be caused by an agent related to the emergent *Arenaviridae*. Nevertheless, these facts were used to arrive at the tentative personal impression that the syndrome was immunopathological and that free virus might not be present in blood or tissues during acute illness. If that were true, the search for the virus(es) should be concentrated on rodents.

Shortly thereafter, Col. Bryce Walton, the US Army parasitologist in charge of the military unit housed in quarters occupied by our NIH team in the Canal Zone, was assigned to Tokyo to direct an infectious disease research program in eastern Asia supported by the Army. He visited each of the then current grantees. A fateful stop was made in Seoul, Korea, to the laboratory of Ho Wang Lee. Dr. Lee had studied medicine in Seoul and then done a Ph.D. thesis at the University of Minnesota under the tutelage of the internationally recognized arbovirologist, William Scherer. Lee was attempting to demonstrate over-wintering of Japanese B Encephalitis virus in snakes. The studies did not prosper, and he frankly did not care for these sometimes-rude candidate hosts. He wanted to try to isolate the virus of Korean hemorrhagic fever (KHF), as HFRS was called in that country. Although American troops stationed in Korea now suffered little disease, mainly because their numbers were quite small, from 300 to 900 cases per year occurred in Korean military and civilian populations stationed or living along the demilitarized zone separating North from South Korea. Walton arranged a trip to Panama for Dr. Lee.

We agreed that rodents rather than patients should be the first targets for etiologic work. Co-cultivation of rodent tissues on monkey kidney-cell cultures and indirect immunofluorescent antibody (IFA) technology were considered as isolation methods. The latter would comprise reacting serial sections of rodent tissues, principally those of reticuloendothelial origin (which were main sites of arenaviral replication) with pooled convalescent serum from typically severe cases of KHF, and, after a rinse, applying an anti-human immunoglobulin tagged with fluorescein isothiocyanate. We used IFA in Panama to measure antibodies to Machupo virus which generally infected cultured cells without causing cytopathic effects. Because Dr. Lee was unfamiliar with this technology and did not have the special microscope needed for it, the former method was chosen. Within two years a pair of distinct viruses had been recovered from *Apodemus agrarius*, which was found to be the most numerous rodent in the area of the country where disease in civilians was paramount. These were both cytopathic agents and turned out to be murine herpes and reoviruses, respectively. Now it was time for immunofluorescence. Fortunately, Dr. Lee did not listen to advice that lymphoid tissues should get first priority. He simply made serial sagittal sections of mice from nose to tail. By late 1975 he had repeatedly found specific staining in tissues of only *Apodemus agrarius*. Viral antigen was predominant in lung, with lesser incidence and intensity in liver and kidney. Before he could report these findings, however, notice arrived from the US Army that his grant was to be terminated. Coded serum samples from convalescent HFRS patients from the Soviet Union and some control sera from individuals who had never ventured to any of the known endemic regions of the globe were sent by Dr. Jordi Casals of Yale University. Dr. Lee called each one correctly; his support was renewed. Shortly thereafter he developed a sensitive diagnostic test for clinical KHF using lung sections from infected wild mice (LEE and LEE 1976).

Attempts to find a laboratory host or cell system for replication of the presumptive virus were initially unfruitful. Mammalogists had been unable to colonize *A. agrarius*, so wild *Apodemus* from the south of the country, where HFRS had not been recognized, were used to make serial passages of lung tissues in order to establish isolation of a replicating virus. In perhaps his most brilliant stroke, Dr. Lee recalled that HFRS had never been reported from Jeju Island located off the south coast of Korea. He captured mice there belonging to the subspecies *jejudoica*, found that none had evidence of infection and used these animals for new isolation attempts, as hosts for study of infection in the rodent, and as substrate for virus titrations in other studies. This subspecies proved even more susceptible to infection than the mainland *Apodemus agrarius coreae*. Infected mouse lung slides were used to diagnose KHF and to examine the pattern of evolution of viral antibodies. Clinically classical illness was confirmed in 113 of 116 instances, but in less than half of those regarded by physicians as mild. Antibodies were present in sera of nearly all patients before or by the 7th day of illness, a first signal that this disease had an immunopathological component (LEE et al. 1978). LEDUC et al. (1990) subsequently confirmed the clinical diagnosis of KHF in 94% of 245 military patients studied in 1951–1954. Not surprisingly, it proved rather difficult to isolate the virus from blood of KHF patients. Only 19 strains were recovered from 105 samples taken during the acute phase of disease (LEE 1982). Whatever the exact nature of this new agent might be, it was not serologically related to any of the other major taxa that caused hemorrhagic fever. Dr. Lee named it Hantaan virus after the small river that passes near the village of Songnaeri, where the prototype strain was obtained.

Now the race to identify and map HFRS infection and viruses was on. Antibodies to Hantaan were quickly found in sera from HFRS patients in Scandinavia, Finland, Russia, China, and both eastern and western Europe (SVEDMYR et al. 1982; LEE 1982). Those in samples from areas where the mild form of the disease occurred were generally of lower titer than in China and eastern Asia (LEE et al. 1979a). Soon workers in Finalnd were able to find florescent antigen in lungs of the principle suspect, *Clethrionomys glareolus*. Reciprocal IFA tests revealed a relationship to, but clear difference from, Hantaan virus (BRUMMER-KORVENKONTIO et al. 1980). Again, this method proved highly specific and sensitive for identification of human infection; in one study LAHDEVIRTA et al. (1984) found that 40 of 41 clinically diagnosed patients had antibodies, and seven less-typical infections also were identified.

In 1964 Japanese physicians had described HFRS among residents of inner city Osaka (Tamura 1964). Autopsy findings in one fatal case were characteristic of the disease (Matsumoto et al. 1964). Exposure of these individuals to rural wild rodents had not occurred. Convalescent sera from some of these patients were tested in Seoul and were found positive (Lee et al. 1979b). This puzzling finding was resolved when several cases of HFRS were observed by Lee in which patients had not left environs of urban Seoul prior to onset of disease. Viral antigen was demonstrated in lungs of rats, both *Rattus rattus* and *R. norvegicus*, and a replicating agent was established in laboratory rats. Based on IFA relationships, Lee et al. (1982) initially considered this virus to be Hantaan, a conclusion that would eventually be revised. Acting on the assumption that infected rats had been distributed to virtually all corners of the globe by ships; this notion was soon amply confirmed and extended to areas far removed from seaports (LeDuc et al. 1984, 1985; Tsai et al. 1985). In addition, human HFRS infections in research personnel at laboratories in several countries were traced to chronic infection of rat colonies (Umenai et al. 1979; Desmyter et al. 1983; Lloyd and Jones 1986). Another virus was recovered from the vole, *Microtus pennsylvanicus* in Maryland, USA, and named Prospect Hill (Lee et al. 1985). This distinct agent has not been associated with acute human disease.

Studies of Hantaan virus infection in the *Apodemus* host were reported by Lee et al. (1981). Viremia was limited to a 3–5 day interval ending about 12 days after infection. At this time, large amounts of virus were present in lung, liver, kidney, and salivary glands. Saliva, urine, and feces contained generous amounts of infectious virus as well, and it was shown that intra-cage transmission occurred after 5 days when susceptible mice of either gender were placed with those inoculated up to about 40 days previously, the interval during which the agent could be detected in saliva and feces. Intercage transmission also occurred, albeit in a more sporadic pattern, during continuous exposure up to 150 days. Virus was detected in urine of infected mice for at least 360 days. Vertical transmission did not occur in limited experiments, a finding reinforced by Childs et al. (1987) in longitudinal field studies of the rat virus done in Baltimore. Finally, horizontal transmission was found to be independent of presence or absence of ectoparasites on the wild mice used in the trials. Arthropods seemed less and less likely to have a role in the host/parasite equation.

4 Virus Characterization and Emergence of a New Genus

It was evident that wild mice were not ideal hosts to use for study of this new group of agents. Although it was known to which families these viruses were not assigned, the number of immunotypes and the tribe to which they did belong, was far from clear. French et al. (1981) finally were able to adapt the prototype 76–118 virus strain from *Apodemus* to growth in cultured cells utilizing A-549, a continuous line of human lung carcinoma cells. Shortly thereafter, electron micrographic pictures of the virus were achieved at virtually the same time in two US laboratories (McCormick et al. 1982; White et al. 1982). The E-6 clone of monkey kidney vero cells used at one laboratory subsequently proved the better substrate for virus detection and replication. This clone, originally derived at the Centers for Disease Control (CDC) and used for quantitative virus assays of the arenavirus, Lassa (Earley and Johnson 1988), has become perhaps the most widely utilized cellular host in vertebrate virology. The pictures, however, proved a bombshell. Hantaan was not an arenavirus; nor a flavivirus or filovirus. It looked like a bunyavirus. The problem was that *Bunyaviridae* is the largest predominately animal virus family known, and all members were firmly established as arthropod-borne.

Once the viruses were cultivated in cells, neutralization tests were employed to document that four distinct agents had been discovered: Hantaan, the rat virus now named Seoul; Puumula, the virus of NE or mild HFRS; and Prospect Hill (Schmaljohn et al. 1985). It was at this time that the concept of generally host-specific parasitism for these viruses began to take hold. This interesting parallel to the rodent-borne *Arenaviridae* has continued to strengthen, even though the immunological bases for host specificity and chronicity almost certainly differ. Monoclonal antibodies were generated to probe the viral relationships in greater detail, and to help generate reagents for new diagnostic tests such as enzyme immunoassays. Molecular characterization of Hantaan virus revealed presence of a tripartite, negative sense genome. Further, it was determined that all Hantaan-related viruses had unique and conserved 3′ terminal nucleotide sequences which were distinct from those of other recognized bunyavirus genera. Hantavirus was proposed and accepted as the name for this new genus, an apparent outlier in the biological world of the larger family (Schmaljohn and Dalrymple 1983). A lethal mouse model was developed for Hantaan virus, and utilized to discover that ribavirin, a broad spectrum anti-viral that had shown efficacy against clinical Lassa fever, was active (Huggins et al. 1986). A ribavirin clinical trial later done in China was successful (Huggins et al. 1991). Inactivated vaccines have been developed and utilized in China and Korea (Zhu et al. 1994; Cho and Howard 1999).

By 1993 several new hantaviruses were under suspicion or had been documented. These included Thottapalyam, an agent recovered from the shrew *Suncus murinus* years previously in India, another from the rat, *Bandicota indica* in Thailand, and a third, Dobrava, from the rodent *Apodemus flavicollis*, in Yugoslavia (Avsic-Zupanc et al. 1992). Thottapalyam virus joined the genus when it was found to have bunyavirus morphology, and the generic 3′ nucleotide signature (Zeller et al. 1989; Xiao et al. 1992). The Thailand virus was shown to be serologically related to Hantaan virus, but distinct at the level of virus neutralization (Chu et al. 1995); and the Dobrava virus was accepted on the basis of serological relationships, together with distinct nucleotide sequences that were unraveled using polymerase chain reaction (PCR) technology already employed to examine relationships among other hantaviruses (Arthur et al. 1992; Avsic-Zupanc et al. 1992; Puthavathama et al. 1992). The stage was set for the next drama.

5 Explosion: New World Viruses, New Disease

Two young Navajo adults living together near Gallup, New Mexico, died within five days of each other after brief but fulminant pulmonary illnesses in May 1993. Richard Malone, a field investigator for the New Mexico State Office of Medical Investigation (OMI) in Albuquerque, learned of this event on the day of the second death (Friday, 14 May). He persuaded the family of the first patient to have an autopsy done to compare findings with those of the patient whom he had already certified as an OMI case. Dr. Patricia McFeeley of the OMI staff did both autopsies, and immediately realized that her findings were similar to those of another case from the Navajo reservation that she had done a month earlier, which left her in doubt as to final diagnosis. Her report was made to the state Department of Health (DOH) that evening. In the meantime, Dr. Bruce Tempest, an experienced internist who had spent many years at the US Public Health Service's Indian Health Service (IHS) hospital in Gallup, where these patients had been treated, also was perplexed. Neither plague nor influenza pneumonia seemed likely diagnoses. Calling colleagues at other IHS health centers in the Four Corners region of the American southwest during the weekend, he obtained descriptions of two other recent cases. His report of five similar severe illnesses went to the DOH on Monday, 17 May. Sister agencies in Arizona, Utah, and Colorado were notified, and clinical and autopsy samples were sent to the CDC. Within three weeks some 24 suspect case-patients had been identified, most of them from New Mexico. Antibodies to hantavirus were detected in human case sera before the month was out, and within another two weeks, the rodent *Peromyscus maniculatus* was identified as the common and sole carrier of the agent (MMWR 1993a,b; Childs et al. 1994). Genetic analysis done on rodent and human tissue, together with serological tests and demonstration of specific viral antigen in capillary endothelium, principally that of lung, demonstrated that a new virus most closely related to Prospect Hill and Puumula viruses was the culprit (Feldmann et al. 1993; Nichol et al. 1993). History matters. Whereas it required more than 25 years to identify the cause of HFRS, the etiology of this new pulmonary syndrome was elucidated in little more than one month. Retrospective studies now confirm occurrence of the disease at least 30 years previously, and it is certain that the virus has been present in the deer mouse long before humans invaded the Americas (Frampton et al. 1995; Zaki et al. 1996).

History now becomes entwined with current affairs. New technology, principally PCR, has provided the "fishing rod" for the detection of so many new and candidate hantaviruses that they will not be listed here by name or number. It suffices to say, the hantavirus pulmonary syndrome (HPS), as it has been labeled, occurs mostly west of the Mississippi River in North America, the virus finally (and ironically) named Sin Nombre (without name) is the leading cause of the disease, and several other hantaviruses vectored by other rodents occasionally induce similar disease (Peters et al. 1999). Ferrer et al. (1998) found high prevalence of hantavirus antibodies among native Americans in the Chaco region of Argentina and Paraguay, and HPS was documented in the Andes region of Argentina and Chile (Levis et al. 1997; Toro et al. 1998). The causative agent has taken the name of the mountains; more importantly, there is strong evidence that person-to-person transmission is significant in the epidemiology of the syndrome (Padula et al. 1998). If confirmed, this would be the first such example for any hantaviral disease. It is also cautionary for medical personnel because mortality in both Americas remains about 50% (Khan et al. 1996; Levis et al. 1997).

In clinical and pathophysiological terms, HPS has many points of similarity to HFRS, but also several distinct attributes (Nolte et al. 1995). In the former disease, the capillary leak is almost never accompanied by escape of erythrocytes, and is more conspicuous in lungs than in other tissues. This deprives clinicians of an early warning sign that is commonly available in HFRS. Further, the dramatic occurrence of this leak, together with a cytokine cascade (Mori et al. 1999) that is the possible trigger for cardiac depression, makes patient management an intensive care crisis. It may render anti-viral therapy moot, because the crisis usually lasts no more than three days, after which the capillary leak disappears as rapidly as it began. What stops the flood of protein and fluid remains to be determined. Hallin et al. (1996) observed that many HPS patients exhibit extremely low cardiac output. These authors emphasize that death in HPS, or more properly HCPS, is caused by cardiac failure rather than by pulmonary edema. Crowley et al. (1998) reported success in treating severe HCPS with extracorporeal membrane oxygenation (ECMO). The pathogenesis of HCPS has been reviewed by Zaki et al. (1995). New information regarding the role of cytolytic T-cells in this syndrome (Ennis et al. 1997) and of major histocompatibility alleles in the severity of nephropathia epidemica (Mustonen et al. 1996, 1998; Kanerva et al. 1998) has identified additional factors that may be important.

The ecology of hantavirus infection in rodent hosts is under study. Long term rodent population dynamics and virus infection rates in the rodents are being correlated with meteorological, land form, and other parameters in global information system maps in order to understand, and eventually predict, complex interactions that lead to increased risk for transmission of virus to humans (Engelthaler et al. 1999; Mills et al. 1999). This effort was stimulated when it became evident that the 1993 outbreak of HPS was associated with very high numbers of *P. maniculatus*, which were thought, in turn, to have occurred in response to El Niño weather patterns during the preceding 18 months (Parmenter et al. 1993). Long-term cycles of rodents and human disease are well documented for HFRS in Scandinavia and Finland, but it is not clear whether these are dependent on global climate cycles (Nystrom 1982).

The widespread geographical distribution of these viruses, the ability of some of them to produce serious, often fatal human disease, and recent technological leaps have opened many new avenues to investigation. It is evident, therefore, that the history of hantaviruses remains in its early chapters.

References

Arnold OH (1944) Die sogenannte Feldnephritis; klinische Studie zur Symptomologie, Pathologie und Artiologie einer akuten diffusion Gefasserkrankung infektioser Genese. Thieme, Leipzig

Arthur RR, Lofts RS, Gomez J, Glass GE, LeDuc JW, Childs JE (1992) Grouping of hantaviruses by small (s) genome segment polymerase chain reaction and amplification of viral RNA from wild-caught rats. Am J Trop Med Hyg 47:210–224

Avsic-Zupanc T, Xiao SY, Stojanovic R, Gligic A, van der Groen G, LeDuc JW (1992) Characterization of Dobrava virus: a hantavirus from Slovenia, Yugoslavia. J Med Virol 38:132–137

Bradford JR (1916) Nephritis in the British troops in Flanders Qland. J Med 9:445–459

Brummer-Korvenkontio M, Vahari A, von Bonsdorff CH, Vuorimies J, Manni T, Penttinen K, Oker-Blom N, Laehdvirta J (1980) Nephropathia epidemica: detection of antigen in bank voles and serologic diagnosis of human infection. J Infect Dis 141:131–134

Casals J, Henderson BE, Hoogstraal H, Johnson KM, Shelokov A (1970) A review of Soviet viral hemorrhagic fevers, 1969. J Infect Dis 122:437–453

Childs JE, Korch GW, Glass GE, LeDuc JW, Shah KV (1987) Epizootiology of hantavirus infections in Baltimore: isolation of a virus from Norway rats, and characteristics of infected rat populations. Am J of Epi 126:55–68

Childs JE, Ksiazek TG, Spiropoulou CF, Krebs JW, Morzunov S, Maupin GO, Gage KL, Rollin PE, Sarisky J, Enscore RE, Frey JK, Peters CJ, Nichol ST (1994) Serologic and genetic identification of *Peromyscus maniculatus* as the primary rodent reservoir for a new hantavirus in the southwestern United States. J Infect Dis 169:1271–1280

Cho HW, Howard CR (1999) Antibody responses in humans to an inactivated hantavirus vaccine (Hantavax). Vaccine 17:2569–2575

Chu Y-K, Jennings G, Schmaljohn A, Elgh F, Hjelle B, Lee H-W, Jenison S, Ksiazek T, Peters CJ, Rollin P, Schmaljohn C (1995) Cross-neutralization of hantaviruses with immune sera from experimentally infected animals and from hemorrhagic fever with renal syndrome and hantavirus pulmonary syndrome patients. J Infect Dis 172:1581–1584

Crowley MR, Katz RW, Kessler R, Simpson SQ, Levy H, Hallin GW, Cappon J, Krahling JB, Wernly J (1998) Successful treatment of adults with severe hantavirus pulmonary syndrome with extracorporeal membrane oxygenation. Crit Care Med F 26:409–414

Desmyter J, Johnson KM, Deckers C, LeDuc JW, Brasseur F, van Ypersele de Strihou C (1983) Laboratory rat associated outbreak of haemorrhagic fever with renal syndrome due to Hantaan-like virus in Belgium. Lancet ii:1445–1448

Earle DP (1954) Symposium on epidemic hemorrhagic fever. Am J Med 16:617–709

Earley EM, Johnson KM (1988) The lineage of the vero, vero 76 and its clone C1008 in the United States. In: Simizu B, Terasima T (eds) Vero cells – origin, properties and biomedical applications. Department of Microbiology, School of Medicine, Chiba University, Japan

Engelthaler DM, Mosley DG, Cheek JE, Levy CE, Komatsu KK, Ettestad P, Davis T, Tanda DT, Miller L, Frampton JW, Porter R, Bryan RT (1999) Climatic and environmental patterns associated with hantavirus pulmonary syndrome, Four Corners region, United States. Emerg Infect Dis 5:87–94

Ennis FA, Cruz J, Spiropoulou CF, Waite D, Peters CJ, Nichol ST, Kariwa H, Koster FT (1997) Hantavirus pulmonary syndrome: CD8+ and CD4+ cytotoxic T lymphocytes to epitopes on Sin Nombre virus nucleocapsid protein isolated during acute illness. Virology 238:380–390

Feldmann H, Sanchez A, Morzunov S, Spiropoulou CF, Rollin PE, Ksiazek TG, Peters CJ, Nichol ST (1993) Utilization of autopsy RNA for the synthesis of the nucleocapsid antigen of a newly recognized virus associated with hantavirus pulmonary syndrome. Virus Res 30:351–367

Ferrer JF, Jonsson CB, Esteban E, Galligan D, Basombrio MA, Peralta-Ramos M, Bharadwaj M, Torrez-Martinez N, Callahan J, Segovia A, Hjelle B (1998) High prevalence of hantavirus infection in Indian communities of the Paraguayan and Argentinean Gran Chaco. Am J Trop Med Hyg 59:438–444

Frampton JW, Lanser S, Nichols CR (1995) Sin Nombre virus infection in 1959. Lancet 346:781–782

French GR, Foulke RS, Brand OA, Eddy GA, Lee HW, Lee PW (1981) Propagation of etiologic agent of Korean hemorrhagic fever in a cultured continuous cell line of human origin. Science 211:1046–1048

Gajdusek DC (1962) Viral hemorrhagic fevers. J Ped 60:841–857

Hallin GW, Simpson SQ, Crowell RE, James DS, Koster FT, Mertz GJ, Levy H (1996) Cardiopulmonary manifestations of hantavirus pulmonary syndrome. Crit Care Med 24:252–258

Hortling H (1944) Leptospirosista muistuttava epidemia Lapissa Sotilaslaak Aikak 19:8–19

Huggins JW, Kim GR, Brand OM, McKee KT Jr (1986) Ribavirin therapy for Hantaan virus infection in suckling mice. J Infect Dis 153:489–497

Huggins JW, Hsiang CM, Cosgriff TM, Guang MY, Smith JI, Wu ZO, LeDuc JW, Zheng ZM, Meegan JM, Wang QN, Oland DD, Gui XE, Gibbs PH, Yuan GH, Zhang TM (1991) Prospective, double-blind, concurrent, placebo-controlled clinical trial of intravenous ribavirin therapy of hemorrhagic fever with renal syndrome. J Infect Dis 164:1119–1127

Ishii S (1942) Studies on Song-go fever. Jap Army Med J 355:1755–1758

Johnson KM, Mackenzie RB, Webb PA, Kuns ML (1965a) Chronic infection of rodents by Machupo virus. Science 150:1618–1619

Johnson KM, Wiebenga NH, Mackenzie RB, Kuns ML, Tauraso NM, Shelokov A, Webb PA, Justines G, Beye HK (1965b) Virus isolations from human cases of hemorrhagic fever in Bolivia. Proc Soc Exp Biol Med 118:113–118

Kanerva M, Vaheri A, Mustonen J, Partanen J (1998) High-producer allele of tumour necrosis factor-alpha is part of the susceptibility MHC haplotype in severe Puumala virus-induced nephropathia epidemica. Scand J Infect Dis 30:532–534

Khan AS, Khabbaz RF, Armstrong LR, Holman RC, Bauer SP, Graber J, Strine T, Miller G, Reef S, Tappero J, Rollin PE, Nichol ST, Zaki SR, Bryan RT, Chapman LE, Peters CJ, Ksiazek TG (1996) Hantavirus pulmonary syndrome: the first 100 US cases. J Infect Dis 173:1297–1303

Kitano M (1944) A study of epidemic hemorrhagic fever. J Manchou Med 19:191–209

Kulagin SM, Fedorova NI, Ketiladze ES (1962) Laboratory outbreak of hemorrhagic fever with renal syndrome (Russian). J Micro Epidemiol Immunol 33:121–126

Kuns ML (1965) Symposium on some aspects of hemorrhagic fevers in the Americas. Epidemiology of Machupo virus infection. II. Ecological and control studies of hemorrhagic fever. Amer J Trop Med Hyg 14:813–816

Lahdevirta J (1971) Nephropathia epidemica in Finland. Ann Clin Res (Finland) 3:1154

Lahdevirta J, Savola J, Brummer-Korvenkontio M, Berndt R, Illikainen R, Vaheri A (1984) Clinical and serological diagnosis of nephropathia epidemica, the mild type of haemorrhagic fever with renal syndrome. J Infect 9:230–238

LeDuc JW, Smith GA, Johnson KM (1984) Hantaan-like viruses from domestic rats captured in the United States. Am J Trop Med Hyg 33:992–998

LeDuc JW, Smith GA, Pinheiro FP, Vasconcelos PFC, Rosa EST, Maiztequi J (1985) Isolation of a Hantaan-related virus from Brazilian rats and serologic evidence of its widespread distribution in South America. J Trop Med Hyg 34:810–815

LeDuc JW, Ksiazek TG, Rossi CA, Dalrymple JM (1990) A retrospective analysis of sera collected by the hemorrhagic fever commission during the Korean conflict. J Infect Dis 162:1182–1184

Lee HW (1982) Korean hemorrhagic fever. Prog Med Virol 28:96–113

Lee HW, Lee PW (1976) Korean hemorrhagic fever. I. Demonstration of causative antigen and antibodies. Korean J Int Med 19:371–384

Lee HW, Lee PW, Johnson KM (1978) Isolation of the etiologic agent of Korean hemorrhagic fever. J Infect Dis 137:298–308

Lee HW, Lee PW, Lahdevirta J, Brummer-Korvenkontio M (1979a) Aetiological relation between Korean haemorrhagic fever and nephropathia epidemica. Lancet 1:186–187

Lee HW, Lee PW, Tamura M, Tamura T, Okuno Y (1979b) Etiological relation between Korean hemorrhagic fever and epidemic hemorrhagic fever in Japan. Biken J 22:41–45

Lee HW, Lee PW, Baek LJ, Song CK, Seong IW (1981) Intraspecific transmission of Hantaan virus, etiologic agent of Korean hemorrhagic fever, in the rodent *Apodemus agrarius*. Am J Trop Med Hyg 30:1106–1112

Lee HW, Baek LJ, Johnson KM (1982) Isolation of hantaan virus, the etiologic agent of Korean hemorrhagic fever, from wild urban rats. J Infect Dis 146:638–644

Lee PW, Amyx HL, Yanagihara R, Gajdusek DC, Goldgarber D, Gibbs CJ Jr (1985) Partial characterization of Prospect Hill virus isolated from meadow voles in the United States. J Infect Dis 152:826–829

Levis S, Rowe JE, Morzunov S, Enria DA, St Jeor S (1997) New hantaviruses causing hantavirus pulmonary syndrome in central Argentina. Lancet 349:998–999

Lloyd G, Jones N (1986) Infection of laboratory workers with hantavirus acquired from immunocytomas propagated in laboratory rats. J Infect 12:117–125

Matsumoto Y, Nishino K, Yagura T (1964) A case report of suspicious epidemic hemorrhagic fever. Biken J 7:95–106

McCormick JB, Sasso DR, Palmer EL, Kiley MP (1982) Morphological identification of the agent of Korean haemorrhagic fever (Hantaan virus) as a member of the bunyaviridae. Lancet 1:765–768

Mills JN, Yates TL, Ksiazek TG, Peters CJ, Childs JE (1999) Long-term studies of hantavirus reservoir populations in the southwestern United States: rationale, potential, and methods. Emerg Infect Dis 5:95–101

MMWR Morb Mortal Wkly Rep (1993a) Outbreak of acute illness – Southwestern United States, 1993. 42:421–424

MMWR Morb Mortal Wkly Rep (1993b) Update: outbreak of hantavirus infection – Southwestern United States, 1993. 42:441–443

Mori M, Rothman AL, Kurane I, Montoya JM, Nolte KB, Norman JE, Waite DC, Koster FT, Ennis FA (1999) High levels of cytokine-producing cells in the lung tissues of patients with fatal hantavirus pulmonary syndrome. J Infect Dis 179:295–302

Mustonen J, Partanen J, Kanerva M, Pietila K, Vapalahti O, Pasternack A, Vaheri A (1998) Association of HLA B27 with benign clinical course of nephropathia epidemica caused by Puumala hantavirus. Scand J Immunol 47:277–279

Mustonen J, Partanen J, Kanerva M, Pietila K, Vapalahti O, Pasternack A, Vaheri A (1996) Genetic susceptibility to severe course of nephropathia epidemica caused by Puumala hantavirus. Kidney Int 49:217–221

Myhrman G (1951) Nephropathia epidemica: a new infectious disease in Northern Scandanavia. Acta Med Scand 140:52–56

Nichol ST, Spiropoulou CF, Morzunov S, Rollin PE, Ksiazek TG, Feldmann H, Sanchez A, Childs J, Zaki S, Peters CJ (1993) Genetic identification of a hantavirus associated with an outbreak of acute respiratory illness. Science 262:914–917

Nolte KB, Feddersen RM, Foucar K, Zaki SR, Koster FT, Madar D, Merlin TL, McFeeley PJ, Umland ET, Zumwalt RE (1995) Hantavirus pulmonary syndrome in the United States: a pathological description of a disease caused by a new agent. Hum Pathol 26:110–120

Nystom K (1982) Epidemiology of HFRS (endemic benign nephropathy – EBN) in Sweden. Scand J Infect Dis Suppl 36:92

Padula PJ, Edelstein A, Miguel SD, Lopez NM, Rossi CM, Rabinovich RD (1998) Hantavirus pulmonary syndrome outbreak in Argentina: molecular evidence for person-to-person transmission of Andes virus. Virology 241:323–30

Parmenter RR, Brunt JW, Moore DI, Ernest S (1993) The hantavirus epidemic in the Southwest: rodent population dynamics and the implications for transmission of hantavirus-associated adult respiratory distress syndrome (HARDS) in the four corners region. Sevilleta LTER publication no. 41. Albuquerque, University of New Mexico, New Mexico

Peters CJ, Simpson GL, Levy H (1999) Spectrum of Hantavirus infection: Hemorrhagic fever with renal syndrome and hantavirus pulmonary syndrome. Annu Rev Med 50:531–545

Puthavathana P, Lee HW, Kang CY (1992) Typing of hantaviruses from five continents by polymerase chain reaction. Virus Res 26:1–14

Schmaljohn CS, Dalrymple JM (1983) Analysis of Hantaan virus RNA: evidence for a new genus of Bunyaviridae. Virology 131:482–491

Schmaljohn CS, Hasty SE, Dalrymple JM, LeDuc JW, Lee HW, von Bonsdorff CH, Brummer-Korvenkontio M, Vaheri A, Tsai TF, Regnery HL, Goldgaber D, Lee PW (1985) Antigenic and genetic properties of viruses linked to hemorrhagic fever with renal syndrome. Science 227:1041–1044

Smadel JE (1953) Epidemic hemorrhagic fever. Am J Pub Hlth 43:1327–1330

Smorodintsev AA, Chudakov VG, Churilov AV (1959) Haemorrhagic nephroso-nephritis. In: Chapter 3: epidemiology. Pergamon Press, New York

Stuhlfauth K (1943) Bericht uber ein neues schlammfieberahnliches krankheitsbild bei Deutschen truppen in Lappland. Dtsch Med Wochenschr 69:439–443

Svedmyr A, Lee PW, Goldgaber D, Yanagihara R, Gajdusek DC, Gibbs CJ, Nystrom K (1982) Antigenic differences between European and East Asian strains of HFRS virus. Scand J Infect Dis Suppl 36:86–87

Tamura M (1964) Occurrence of epidemic hemorrhagic fever in Osaka City: first cases found in Japan with characteristic feature of marked proteinuria. Biken J 7:79–94

Toro J, Vega JD, Khan AS, Mills JN, Padula P, Terry W, Yadon Z, Valderrama R, Ellis BA, Pavletic C, Cerda R, Zaki S, Shieh WJ, Meyer R, Tapia M, Mansilla C, Baro M, Vergara JA, Concha M, Calderon G, Enria D, Peters CJ, Ksiazek TG (1998) An outbreak of hantavirus pulmonary syndrome, Chile, 1997. Emerg Infect Dis 4:687–694

Traub R, Hertig M, Lawrence WH, Harriss TT (1954) Potential vectors and reservoirs of hemorrhagic fever in Korea. Am J Hyg 59:291–305

Trencseni T, Keleti B (1971) Clinical aspects and epidemiology of haemorrhagic fever with renal syndrome. Akademiai Kiado, Budapest

Tsai TF, Bauer SP, Sasso DR, Whitfield SG, McCormick JB, Caraway TC, McFarland L, Bradford H, Kurata T (1985) Serological and virological evidence of a Hantaan virus-related enzootic in the United States. J Infect Dis 152:126–144

Umenai T, Lee PW, Toyoda T, Yoshinaga K, Horiuchi T, Lee HW, Saito T, Hongo M, Nobunaga T, Ishida N (1979) Korean haemorrhagic fever in staff in an animal laboratory. Lancet 1:1314–1316

White JD, Shirey FG, French GR, Huggins JW, Brand OM, Lee HW (1982) Hantaan virus, aetiological agent of Korean haemorrhagic fever, has bunyaviridae-like morphology. Lancet 1:768–771

Xiao S-Y, Chu Y-K, Knauert FIK, Lofts R, Dalrymple JM, LeDuc JW (1992) Comparison of hantavirus isolates using a genus reactive primer pair polymerase chain reaction. J Gen Virol 73:567–573

Yankovski AK, Povalishina TP (1967) Hemorrhagic fever with a renal syndrome caused by aerial contamination. Mil Med J (Russian) 7:66–69

Zaki SR, Greer PW, Coffield LM, Goldsmith CS, Nolte KB, Foucar K, Feddersen RM, Zumwalt RE, Miller GL, Khan AS, et al. (1995) Hantavirus pulmonary syndrome. Pathogenesis of an emerging infectious disease. Am J Pathol 146:552–579

Zaki SR, Khan AS, Goodman RA, Armstrong LR, Greer PW, Coffield LM, Ksiazek TG, Rollin PE, Peters CJ, Khabbaz RF (1996) Retrospective diagnosis of hantavirus pulmonary syndrome, 1978–1993: implications for emerging infectious diseases. Arch Pathol Lab Med 120:134–139

Zeller HG, Karabatsos N, Calisher CH, Digoutte J-P, Cropp CB, Murphy FA, Shope RE (1989) Electron microscopic and antigenic studies of uncharacterized viruses. II. Evidence suggesting the placement of viruses in the family Bunyaviridae. Arch Virol 108:211–227

Zhu ZY, Tang HY, Li YJ, Weng JQ, Yu YX, Zeng RF (1994) Investigation on inactivated epidemic hemorrhagic fever tissue culture vaccine in humans. Chin Med J 107:167–170

Replication of Hantaviruses

C.B. JONSSON* and C.S. SCHMALJOHN**

1 Introduction

Hantaviruses have a genome consisting of three segments of negative-sense, single-strand RNA (reviewed in SCHMALJOHN 1996a). The large (L), medium (M), and small (S) genomic segments, or vRNAs, code for the viral RNA-dependent RNA polymerase (RdRp), envelope glycoproteins (G1 and G2), and nucleoprotein (N), respectively. Nonstructural proteins have not been described for hantaviruses. One of the first molecular features found to distinguish hantaviruses from other viruses in the family *Bunyaviridae* is the sequence of their conserved, complementary terminal nucleotides on the L, M, and S segments (SCHMALJOHN and DALRYMPLE 1983). This characteristic, and the absence of serological cross-reactivity among other members of the family, were the bases for the proposal to establish the *Hantavirus* genus in the family *Bunyaviridae* in 1986.

* Department of Chemistry and Biochemistry, New Mexico State University, Box 30001, New Mexico State University, Las Cruces, NM 88003, USA
** United States Army Medical Research Institute of Infectious Diseases, Fort Detrick, Frederick, MD 21702, USA

Several reviews detailing the basic molecular properties and coding strategies of hantaviruses exist (Plyusnin et al. 1996; Schmaljohn 1996a,b, 1998). In this chapter, we will focus on current information regarding molecular and biochemical characteristics of hantaviruses that might be important for viral replication and assembly. Where possible, we will compare and contrast the properties of hantaviruses to those of other negative-strand RNA viruses, particularly those of other viruses in the family *Bunyaviridae*.

2 Components of Replication

The minimal components of replication are the RdRp, N, and the viral genomic and antigenomic RNA templates. In vitro reconstitution of the replicase activity of Hantaan virus, with purified virions, demonstrates the presence of the RdRp in the virion (Schmaljohn and Dalrymple 1983). After entry of the virion into the cytoplasm, the RdRp initiates viral RNA synthesis from the genomic L, M, and S RNAs (vRNA) and the antigenomic, or complementary RNAs (cRNA), which are complexed with N throughout transcription and replication (reviewed in Kolakofsky and Hacker 1991). No information is available concerning possible host factors involved in hantaviral replication.

2.1 The Replication Templates

Isopycnic and rate zonal centrifugation studies of detergent-disrupted Hantaan virions reveal that the tripartite genome forms three separate ribonucleocapsids (Schmaljohn et al. 1983). Both the ribonucleocapsids and the metal-dependent RdRp activity of Hantaan virus are sensitive to ribonuclease A, suggesting that N does not completely protect the RNA of the transcription/replication templates (Schmaljohn and Dalrymple 1983). Clearly, the RdRp must associate with the ribonucleocapsids for transcription, but the manner by which this occurs is unknown.

The presence of the highly conserved, complementary, 3′ and 5′ terminal nucleotides on all three genome segments of hantaviruses and other viruses in the family suggests that they, or the panhandle structures that they are predicted to form, may be important in viral replication. Although direct evidence for panhandle formation of the RNAs of hantaviruses has not been obtained, such structures were visualized by electron microscopy for other viruses in the family (Hewlett et al. 1977). For the bunyavirus, LaCrosse virus, the terminal nucleotides can base pair to form panhandle structures, even when the vRNA and cRNA are complexed with N (Raju and Kolakofsky 1989). Also for Bunyamwera virus, studies with a reporter system demonstrated that both the 3′ and 5′ terminal nucleotides that make up the panhandle are required for efficient polymerase activity (DUNN et al. 1995). In those studies removing any of the 5′ nucleotides or removing two or more 3′ terminal nucleotides greatly reduced reporter activity. Similar studies with another negative-strand RNA virus, influenza virus, demonstrated that the panhandle RNA sequences are sufficient to signal transcription initiation (LUYTES et al. 1989; SEONG and BROWNLEE 1992; FODOR et al. 1994).

Bioinformatic analysis of the 1696 nucleotides of the Hantaan virus S segment vRNA, by using the RNA secondary structure prediction program MFOLD, version 3.0, (MATHEWS et al. 1999; ZUCKER et al. 1999) suggests that base-pairing might extend beyond the terminal complementary nucleotides presumed to form the panhandle. Such RNA modeling clearly shows the predicted panhandle structure and shows one major foldback structure with several interspersed stem-loop structures along the backbone (Fig. 1). Of course, this prediction does not take into account the effects of complexed N, which likely would influence the structure of the RNA as well as affect transcription and replication.

Although direct examination of the role of the terminal nucleotides in hantavirus replication awaits a reverse genetics system, an indirect measure of their importance is suggested by the results of a study with Seoul virus. In that study, the terminal nucleotides of the gene segments were characterized at various times after establishment of persistently infected cell cultures. As persistence progressed, terminal deletions of the L, M, and S segments, but particularly the L segment, correlated with decreased levels of replication (MEYER and SCHMALJOHN 2000a). These data are consistent with a role for the panhandle-forming region of the gene segments as transcription recognition sites.

2.2 The RdRp

The RdRps of viruses in the family *Bunyaviridae*, which range in size from about 240kDa for phleboviruses and hantaviruses, to more than 450kDa for nairoviruses, are among the largest of known polymerases (reviewed in ELLIOTT 1996). To mediate both transcription and replication, the RdRp must perform numerous unique, enzymatic functions. As detailed below, the RdRp must have endonuclease, transcriptase, replicase, and possibly, RNA helicase activity. Additional biochemical functions, such as proofreading, remain theoretical possibilities; however, the mutation frequency reported for hantaviruses, of approximately 1×10^{-3} (PLYUSNIN et al. 1996), is similar to other viral RNA polymerases, which strongly suggests the absence of proofreading or repair mechanisms (DOMINGO and HOLLAND 1994). To mediate the activities listed, the RdRp should contain regions or domains for binding the host mRNA cap (and perhaps the associated RNA) and the genomic RNA template. How the RdRp accomplishes these various functions is not known. One possibility is that different structural conformations for the RdRp are responsible for guiding transcription and replication functions. The shift toward one conformation of the polymerase with its template might be triggered by interaction of the polymerase with other viral components, such as N and/or the viral RNA products, with or without the influence of cellular

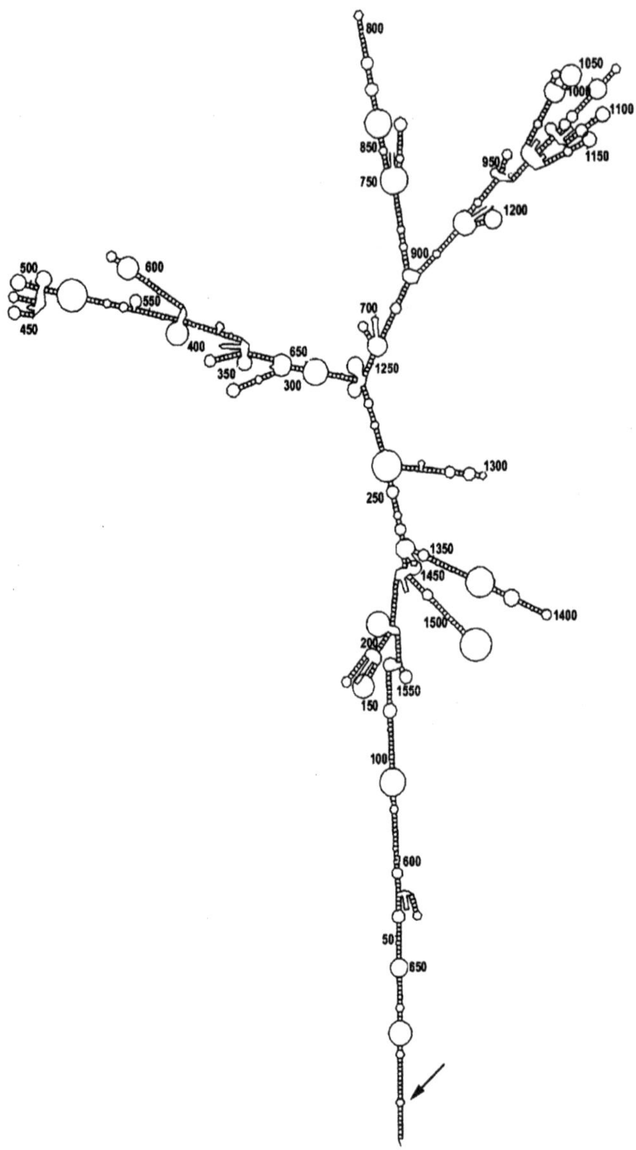

Fig. 1. Predicted secondary structure of the Hantaan S segment vRNA. The MFOLD program version 3.0 was used to generate a model of the Hantaan S segment vRNA using a linear RNA sequence with a fixed folding temperature of 37°C at 1M NaCl with no divalent ions. (Figure and modeling generated by Liza Gonzalez, New Mexico State University, Program in Molecular Biology)

components such as host mRNA. Precedent for equilibrium between two or more conformations of enzymes, including polymerases, is well documented in the literature. For example the "shuttle mechanism" for DNA polymerases suggests that a movement of the 3′ end of the nascent DNA strand between the exonuclease active site and the polymerase active site is at equilibrium and occurs faster than the incorporation of the next nucleotide (STEITZ 1999). In this model, a mismatch shifts the equilibrium toward the editing mode; i.e., into the exonuclease active site. The influenza virus polymerase has also been shown to undergo conformational changes when it interacts with the viral vRNA 5′ and 3′ end sequences (LI et al. 1998). The 5′ and 3′ end sequences bind sequentially to different sites in the PB1 protein. Binding the 5′-end sequences creates an allosteric alteration, which permits the binding of the 3′ end of the viral genome and 5′-capped ends of host RNA. This binding induces in a second allosteric alteration that activates the endonuclease function.

Although molecular and biochemical information on the structure and function relationships of the RdRps of viruses in the family *Bunyaviridae* are not available, studies with other polymerases provide clues to possible regions of the hantaviral RdRp associated with its various activities. The crystal structure analyses of a number of polymerases show they share common architectural features (STEITZ 1999). The overall shape of the polymerase is compared to that of a right hand, with "fingers", "palm", and "thumb" domains. The first X-ray crystal structure of a viral RdRp, the poliovirus polymerase, shows that structural conservation extends to RdRps, although structural differences were observed in the fingers and thumb (HANSEN et al. 1997). Amino acid alignments of these well-characterized polymerase domains, with polymerase sequences that are not characterized, distinguish areas where these domains should occur. Such an alignment of the amino acid sequences predicted for RNA virus polymerases reveals four major conserved motifs in the palm region, designated as A–D, that are shared by unsegmented and segmented negative-strand RNA viruses, positive strand viruses, retrotransposons, and retroviruses (POCH et al. 1989; MULLER et al. 1994; O'REILLY and KAO 1998). In addition to motifs A–D, a region unique to RdRp and reverse transcriptase (RT), motif E, follows motif D in the palm (POCH et al. 1989). The fingers region lies to the amino-terminal side of the palm and includes motif A, while the thumb region follows the palm domain. Alignment of the amino acids spanning this region of the RdRp of several members of the family *Bunyaviridae* show the location of the A–D in these polymerases (Fig. 2). Motifs A and C are involved in magnesium coordination, while motifs A and B may be involved in sugar selection. Motif C contains the catalytic core residues or the XDD box that is essential for catalytic activity and results in a loss of polymerase activity when mutated (SANKAR and PORTER 1992; O'REILLY and KAO 1998). For Bunyamwera virus, site-directed mutagenesis studies have demonstrated the importance of the amino acids in this conserved region of the RdRp motif C, SDD, which spans amino acids 1163–5 (JIN and ELLIOTT 1992). Changes in S at position 1163 and in D at position 1165 resulted in loss of viral RNA synthesis in a transient assay containing expressed RdRp and authentic Bunyamwera nucleocapsids. Similar loss of transcriptase activity was observed when substitutions for the highly conserved D at position 1037, in motif A, were introduced. Site-directed mutageneses of several nonconserved amino acids were also made, but these had no effect on polymerase activity.

```
                       981
Peanut Bud Necrosis    .......... .LSADQSKWS ASDLT.YKYI ..<40>
Tomato Spotted Wilt    .......... .LSADQSKWS ASGLTTYKYV ..<40>
La Crosse              SLGKAKGLKM EINADMSKWS AQDVF.YKYF ..<40>
Bunyamwera             .....KALKL EINADMSKWS AQDVF.YKYF ..<40>
Sin nombre             ...KFKRRLM YVSADATKWS PGDNSAKFKR ..<40>
Puumala                ...KLKRKLM YVSADATKWS PGDNSAKFRR ..<40>
Hantaan                ...RMKRKLM YVSADATKWS PGDNSAKFRR ..<40>
Seoul                  ...RMKRKLM YVSADATKWS PGDNSAKFRR ..<40>
CONSENSUS              SLGKMKRLKM EISADQSKWS ASDLTAYKYF ..<40>
                                   Motif A
                       1051                                               1100
Peanut Bud Necrosis    KRRQTEYGSY GTALSVLADN LETNTFPVSM NWLQGNLNYL SSVYHSCAMM
Tomato Spotted Wilt    RKAQQTFGEN ETAIGLLTKG LTTNTYPVSM NWLQGNLNYL SSVYHSCAMK
La Crosse              LDQKVAY..Q NDIIATMTNQ LNSNTVLIKR NWLQGNFNYT SSYVHSCAMS
Bunyamwera             LDQKRPY..N DDLILEMTNG LNYNYVQIKR NWLQGNFNYI SSYVHSCAML
Sin nombre             DSMDEHSEAV QDFLDFFKGG VSAT...VKG NWLQGNLNKC SSLFGAAVSL
Puumala                DNMGELSDEV LDFLSFFPNK VSAS...IKG NWLQGNLNKC SSLFGAAVSL
Hantaan                DSMESLDPHI KQFLDFFPDG HHGE...VKG NWLQGNLNKC SSLFGVAMSL
Seoul                  DSMDTYEPHV RDFLNFFPDG HHGE...VRG NWLQGNLNKC SSLFGVAMSL
CONSENSUS              KDQQQTYGEN NDFISFMTNG LNTNTYPVKR NWLQGNLNYC SSYYHSCAMM
                                                            Motif B
                       1101                                               1150
Peanut Bud Necrosis    GYVKAMQ. K MKDYDFTIRW  MVHSDDNAT  .......        SIVVRGDIKK
Tomato Spotted Wilt    AYHNTLE..C YKNCDFQTRW  IVHSDDNAT   .......       SLIASGEVDK
La Crosse              VYKEILKEAI TLLDGSILVN SLVHSDDNQT    ......       SITI......
Bunyamwera             VYKDILKECM KLLDGDCLIN SMVHSDDNQT   ......        SLAI......
Sin nombre             LFRRIWAELF PELECFFEFA   HHSDDALFIY GYLEPEDDGT DWFLYVSQQI
Puumala                LFKRVWAKLY PELECFFEFA   HHSDDALFIY GYLEPVDDGT EWFQYVTQQI
Hantaan                LFKQVWTNLF PELDCIFEFA   HHSDDALFIY GYLEPVDDGT DWFLFVSQQI
Seoul                  LFKEIWTRLF PELDCFFEFA   HHSDDALFIY GYLEPADDGT DWFLFVSQQI
CONSENSUS              VYKEVLKECF KMLDCDFQFA   HHSDDALFIN WMVHSDDNAT SLVIYGDVQK
                                                 Motif C
                       1151                                               1200
Peanut Bud Necrosis    LLSSFNCSSL SELLFRSI.. ..QSHFKSYC ITLNPKKSYA SESEVEFISE
Tomato Spotted Wilt    MLTDFSSSSL PEMLFRSI.. ..EAHFKSFC ITLNPKKSYA SSSEVEFISE
La Crosse              VQDKMENDKI IDFAMKEF.. ..ERACLTFG CQANMKKTYV TNCIKEFVSL
Bunyamwera             IQNKVSDQIV IQYAANTF.. ..ESVCLTFG CQANMKKTYI THTCKEFVSL
Sin nombre             QAGNYHWHAV NQEMWKSMFN LHEHLLLMGS IKVSPKKTTV SPTNAEFLST
Puumala                QAGNFHWHAV NQEMWKSMFN LHEHILLMGS IKISPKKTTV SPTNAEFLST
Hantaan                QAGHLHWFSV NTEMWKSMFN LHEHILLLGS IKISPKKTTV SPTNAEFLST
Seoul                  QAGKLHWFNV NTEMWKSMFN LHEHILLLGS IKISPKKTTL SPTNAEFLST
CONSENSUS              MQNKFHWSSI NEYMWKSFFN LHEHHCLTFC CQLNPKKTYV SNTEKEFIST
                                                         Motif D     Motif E
```

Fig. 2. Amino acid sequence alignment of the region spanning Motifs A–E in the *Bunyaviridae* RdRps. Highly conserved amino acids are highlighted in bold. This region was abstracted from a full-length alignment of the RdRps by the ProDom protein domain database (http://protein.toulouse.inra.fr/prodom.html), which compiles alignments of homologous domains. RdRp amino acids sequences are from the following viruses: Peanut bud necrosis (TrEMBL ID O56078), tomato spotted wilt, La Crosse (TrEMBL ID Q82998), Bunyamwera (Swiss Prot ID P204470), Sin Nombre (TrEMBL ID Q89709), Puumala (Swiss Prot ID P27176), Hantaan (Swiss Prot ID P23456), and Seoul (Swiss Prot ID P27314) viruses

2.3 The Nucleocapsid Protein

The vRNA and cRNA replication templates are selectively encapsidated by N during replication by the viral RdRp. Few biochemical studies have addressed the interactions of N and the viral RNAs during replication. We know very little regarding either the region(s) of N that interacts with RNA or the regions of RNA that impart encapsidation specificity.

Studies to identify the region(s) of N important for encapsidation may have been hampered, in part, by difficulties in obtaining sufficient quantities of soluble N to perform biochemical studies. Expression and purification of recombinant N protein in bacterial cells shows the Sin Nombre N protein to have a greater solubility than the Hantaan virus N protein (Severson et al. 1999). The insolubility of

Hantaan virus N when expressed in bacteria (and the relatively less- but still-apparent insolubility of the N of other hantaviruses) may be attributed to one or more intrinsic properties such as their hydrophobicity and natural propensity for aggregation or formation of higher ordered complexes. The solubility problems associated with purifying hantaviral N may also be due to improper folding of the protein in *Escherichia coli* because of the slow rate of proline isomerization as well as the improper reduction or oxidation of cysteine residues (which may impart secondary or higher ordered structures by forming intra- and intermolecular bonds; Hughes and Stock, in press). In addition to overcoming possible solubility problems, purifying N for biochemical studies requires the absence of contaminating ribonucleases, nucleic acid, and RNA-binding proteins.

Despite these obstacles, two in vitro studies addressed the interaction of N with hantaviral RNAs, by analysis of the binding affinity of bacterially expressed and purified hantaviral N with viral and nonviral RNA. In one study, efforts to demonstrate a specific interaction failed (Gott et al. 1993), although binding of viral RNA to filter-immobilized N was observed. In another study, a filter-binding assay demonstrated a moderate binding preference of Hantaan virus N for the S-segment vRNA compared to an RNA comprising only the open reading frame of the S segment, and a strong preference for vRNA compared to nonspecific RNA (Severson et al. 1999). The apparent dissociation constant measured for the interaction between Hantaan virus N and vRNA was 53nM and 35nM in the presence and absence of 1mM $MgCl_2$, respectively (Severson et al. 1999). These values compare similarly to the binding constants reported for ribosomal proteins as well as other RNA binding proteins, which typically range from 10^7 to $10^9 M^{-1}$ (Draper 1995).

Although the mechanism of the specific interaction of N with vRNA and cRNA during viral replication is still unclear, it is likely that sequences or structures present in the 5′ ends of the RNA molecules are involved, perhaps by providing a point of nucleation for subsequent encapsidation of the entire genomic or antigenomic segment. This type of model would necessarily invoke protein–protein interactions that could also involve positive, cooperative interaction of multiple copies of N during the encapsidation of the vRNA and cRNA; that is the binding of the first N protomer to the viral RNA could affect subsequent binding of other N protomers with the RNA molecule. Self-association of N was reported for the tospovirus, tomato spotted wilt virus (Uhrig et al. 1999). In those studies, deletion analysis revealed two binding regions in N, located at amino and carboxy termini. Two highly conserved phenylalanine residues in the carboxy terminal region were found critical for the multimerization (Uhrig et al. 1999). For Hantaan virus, baculovirus expression of S-induced nucleocapsid-like structures, and expression of both M and S by recombinant vaccinia viruses resulted in assembly of virus-like particles (Betenbaugh et al. 1995). In both the tospovirus and hantavirus experiments, viral RNA was clearly not required for assembly; however, neither study eliminated the potential contribution of nonspecific interactions with non-viral nucleic acids.

A model for encapsidation could be envisioned to have both specific and nonspecific interactions of N with the vRNA and cRNA. For example, initially, a specific interaction of N at a sequence and/or structure in the 5′ end could be followed by protein–protein interactions, which could drive the nonspecific binding of the RNA template. This model implies that mRNA, which does not contain the 5′ end sequences, would only interact nonspecifically with N. Experimental evidence for this hypothesis was provided in filter-binding studies that compared the binding of Hantaan virus N with vRNA and mRNA (SEVERSON et al. 1999). The dissociation constant measured for the interaction of N and the RNA comprising the open reading frame of S, in the presence of 1mM $MgCl_2$, was 270nM, a nearly five-fold increase from the interaction with Hantaan virus S vRNA. The increase in the dissociation constant reflects a much weaker interaction of Hantaan virus N with this substrate.

Deletion mapping studies, in which fragments of Hantaan virus or Puumala virus N were refolded after transfer to nitrocellulose, and incubated in situ with viral RNAs, indicated a RNA binding domain in the carboxyl-terminal 93 amino acids of N. This region, however, showed no apparent specificity for viral RNA (GOTT et al. 1993). Motifs common to other RNA binding proteins (SIOMI and DREYFUSS 1997) have not been identified in hantaviral N proteins. As more powerful bioinformatic tools are developed, relationships among RNA binding proteins and their respective motifs may become more apparent. Two of the newest tools, blocks database analysis and LAMA (local alignment of multiple alignments), provide methods to screen for amino acid similarity between protein families by using multiply aligned sequences rather than individual amino acid sequences.

The blocks database[1] (HENIKOFF and HENIKOFF 1994) contains multiply aligned, ungapped segments documented in the Prosite[2] (HOFMANN et al. 1999), Pfam[3], or other similar databases. The alignments are used by the Block Maker program to define the most highly conserved regions for a protein family. Each protein family profile is tracked by its database designation, such as PF00846, which is the Pfam database designation for the hantavirus N proteins. Users can also create their own alignments and use these to generate blocks of proteins not in the database. Bioinformatic analysis of hantaviral N amino acid sequences by the blocks database reveals nine distinct, conserved regions (blocks), which are designated PF00846 A–I (Fig. 3). The blocks map (available in the blocks database) was constructed from multiple alignment of the N amino acids of Hantaan (Swiss Prot P05133), Seoul (Swiss Prot ID P17881), Prospect Hill (Swiss Prot ID P22047), and Puumala (Swiss Prot ID P41268) viruses. The locations of the individual blocks are presented in Fig. 3 in the context of additional hantavirus sequences.

Using LAMA to search the blocks database for sequences similar to those of the aligned hantavirus N sequences reveals a similarity between the L27 ribosomal protein block, BL00831B, which contains the L27 ribosomal protein signature, and the hantavirus N protein Block C (PF00846 C; C. Jonsson, unpublished data).

---

[1] The blocks database, http://blocks.fhcrc.org/blocks, is maintained at the Fred Hutchinson Cancer Research Center in Seattle, Washington, USA.
[2] The Prosite database, http://www.expasy.ch/prosite/, is maintained by the Swiss Institute of Bioinformatics 1211 Genève 4, Switzerland.
[3] The Pfam database, http://pfam.wustl.edu/, is maintained by Washington University in St. Louis.

```
              Block A                                                                 Block B
        1             15 16            30 31            45 46             60 61            75 76             90
TOP     MSNLKDIQDEITRYE QQLIVARQKLRDAEK TVEVDPDDVNKNTLQ ARRQTVSALEDKLAD FKRQLADHVSRQKMD EKPSDPTGIEPDDHL
PUU     MSDLTDIQEEITRHE QQLVVARQKLKDAER AVEVDPDDVNKNTLQ ARQQTVSALEDKLAD YKRRMADAVSRKKMD TKPTDPTGIEPDDHL
ISLA    MSQLREIQEEITRHE QQLVIARQKLKDAEK TVEVDPDDVNKSTLQ SRRAAVSTLEDKLAD FKRQLADAISRQKMD DKPADPTGLEPDDHL
PH      MSQLRETQEEITRHE QQLVIARQKLKEAER TVEVDPDDVNKSTLQ SRRSAVSTLEDKLAE FKRQLADVISRQKMD EKPVDPTGIELDDHL
SN      MSTLKEVQDNITLHE QQLVTTRQKLKDAER AVELDPDDVNKSTLQ SRRAAVSALETKLGE LKRELADLIAAQKLA SKPVDPTGIEPDDHL
HTN     MATMEELQREINAHE GQLVIARQKVRDAEK QYEKDPDELNKRTLT DREGVAVSIQAKIDE LKRQLADRIATGKNL GKEQDPTGVEPGDHL

                                                                         Block C
        91           105 106           120 121           135 136           150 151           165 166           180
TOP     KERSSLRYGNVLDVN AIDIEEPSGQTADWF TIGVYIVSFTLPIIL KALYMLSTRGRQTVK ENKGTRIRFKDDSSF EDVNGIRRPKHLYVS
PUU     KERSSLRYGNVLDVN AIDIEEPSGQTADWY TIGVYVIGFTIPIIL KALYMLSTRGRQTVK ENKGTRIRFKDDTSF EDINGIRRPKHLYVS
ISLA    KERSSLRYGNVLDVN SIDIDEPSGQTADWL QIGSYIISFALPIIL KALYMLSTRGRQTVK ENKGTRIRFKDDSSF EEINGIRRPKHLYVP
PH      KERSSLQYGNVLDVN SIDIEEPSGQTADWL KIGSYIIEFALPIIL KALHMLSTRGRQTVK ENKGTRIRFKDDSSY EDVNGIRRPKHLYVS
SN      KEKSSLRYGNVLDVN SIDLEEPSGQTADWK SIGLYILSFALPIIL KALYMLSTRGRQTIK ENKGTRIRFKDDSSY EEVNGIRKPRHLYVS
HTN     KERSMLSYGNVLDLN HLDIDEPTGQTADWL SIIVYLTSFVVPILL KALYMLTTRGRQTTK DNKGTRIRFKDDSSF EDVNGIRKPKHLYVS

             Block D                                    Block E                                     Block F
        181          195 196           210 211           225 226           240 241           255 256           270
TOP     MPTAQSTMKAEELTP GRFRTIVCGLFPAQI QARNIMSPVMGVIGF SFFVKDWPERIKNFL ESKCPFIKPEVKPGA PAGEADFLSRNQIYF
PUU     MPTAQSTMKAEELTP GRFRTIVCGLFPTQI QVRNIMSPVMGVIGF SFFVKDWPERIRDFM EKESPFIKPEVKPGT PAQEIEFLKRNRVYF
ISLA    MPTAQSTMKADELTP GRFRTIVCGLFPAQI MNRNIISPVMGVIGF AFFVKDWEDKIQEFL RKQCPFLKSEARPGQ SAEEAAILSSNRAYL
PH      MPTAQSTMKAEELTP GRFRTIVCGLFPAQI MARNIISPVMGVIGF AFFVKDWADKVKAFL DQKCPFLKAEPRPGQ PAGEAEFLSSIRAYL
SN      MPTAQSTMKADEITP GRFRTIACGLFPAQV KARNIISPVMGVIGF SFFVKDWMERIDEFL AARCPFLPEQKDP-- --RD-AALATNRAYF
HTN     LPNAQSSMKAEEITP GRYRTAVCGLYPAQI KARQMISPVMSVIGF LALAKDWSDRIEQWL IEPCKLLPDTAAVSL LGGP----ATNRDYL

                                                                   Block G
        271          285 286           300 301           315 316           330 331           345 346           360
TOP     MRRQEVLEDNHIPDI DKLLEYASSGDPTAP DSIESPYAPWVFACA PDRCPPTCIYIAGMA ELGAFFSILQDMRNT IMASKTVGTAEEKLK
PUU     MTRQDVLDKNHVADI DKLIDYAASGDPTSP DDIESPNAPWVFACA PDRCPPTCIYVAGMA ELGAFFSILQDMRNT IMASKTVGTAEEKLK
ISLA    LNRQSVLDESHLPDI DALLEYAASGDPTLP DSLETPHAAWVFACA PDRCPPTCIYVAGMA ELGAFFSILQDMRNT IMASKTVGTADEKLK
PH      MNRQAVLDETHLPDI DALVELAASGDPTLP DSLENPHAAWVFACA PDRCPPTCIYIAGMA ELGAFFAILQDMRNT IMASKTVGTAEEKLK
SN      ITRQMQVDESKVSDI EDLIADARAESATIF ADIATPHSVWVFACA PDRCPPTALYVAGMP ELGAFFGILQDMRNT IMASKSVGTSEEKLK
HTN     RQRQVALGNMETKES KAIRQHAEAAGCSMI EDIESPSSIWVFAGA PDRCPPTCLFIAGIA ELGAFFSILQDMRNT IMASKTVGTSEEKLR

                      Block H                             Block I
        361          375 376           390 391           405 406           420 421
TOP     KKSSFYQSYLRRTQS MGIQLDQRIILLYMV EWGKEMVDHFHLGDD MDPDLRNLAQSLIDQ KVKEISNQEPLKI 433
PUU     KKSSFYQSYLRRTQS MGIQLDQRIILLYML EWGKEMVDHFHLGDD MDPELRGLAQSLIDQ KVKEISNQEPLKI 433
ISLA    KKSAFYQSYLRRTQS MGIQLDQRIILLYMI EWGTDIVNHFHLGDD MDPELRNLAQALIDA KVKEISNQEPLKI 433
PH      KKSAFYQSYLRRTQS MGIQLDQRIILMYMI EWGNEVVNHFHLGDD MDPELRQLAQALIDQ KVKEISNQEPLKI 433
SN      KKSAFYQSYLRRTQS MGIQLDQKIIILYMS HWGREAVNHFHLGDD MDPELRELAQTLVDI KVREISNQEPLKL 428
HTN     KKSSFYQSYLRRTQS MGIQLGQRIIVLFMV AWGKEAVDNFHLGDD MDPELRTLAQSLIDV KVKEISNQEPLKL 429
```

Fig. 3. Amino acid sequence alignment of several hantavirus nucleocapsid proteins. A multiple sequence alignment of several hantavirus N proteins sequences was made with ClustalW using the following viral sequences: Topografov, TOP (GenBank ID AJ011646), Puumala, PUU (GenBank ID Z84204), Isla Vista, ISLA (GenBank ID U31535), Prospect Hill, PH (GenBank ID M34011), Sin Nombre (CC107), SN (GenBank ID AAA68038.1), and Hantaan, HTN (GenBank ID AAA43837.1). The regions outlined correspond to the blocks predicted by the blocks database

The L27 signature, G-x-[LIVM](2)-x-R-Q-R-G-x(5)-G, is shared among the eubacterial L27, algal and plant chloroplast L27, and the yeast mitochondrial YmL2 proteins. L27, which is part of the 50S ribosomal subunit, is thought to contribute to peptidyl transferase activity by mediating the placement of the acceptor end of the tRNA in the active site (Wower et al. 1998). A similar analysis of the arenavirus N protein reveals a region of amino acid similarity between an arenavirus block, designated PF00843B, and the ribosomal L13 protein, BL00783 A (C. Jonsson, unpublished data). The ribosomal L13 protein is also located in the large ribosomal subunit and is shared among the groups listed for L27 as well as mammals. The presence of these motifs in the nucleocapsid proteins of both arenaviruses and hantaviruses suggests that N may have evolved from a component of the primitive RNA translation machinery. Experiments to ascertain whether these motifs confer any structural and/or functional character to N have not been performed, but may prove interesting.

3 The Replication Pathway

3.1 Transcription vs Replication

For all negative-strand RNA viruses, an initial short burst of transcription (primary transcription) by the virion-associated RdRp occurs shortly after virion entry and uncoating in the cytoplasm of host cells. Sometime after primary transcription, an as-yet uncharacterized signal prompts the RdRp to begin replication; i.e., synthesis of the antigenomic cRNA, which will in turn be used as templates for production of vRNA (Fig. 4). To produce vRNA and cRNA, the transcription termination signals recognized during synthesis of mRNA, which result in 3′-truncated molecules, are ignored, and cRNA or vRNA transcripts that are exact complementary copies of the templates are produced. Synthesis of vRNA early in infection provides additional templates for mRNA production and, in later stages, serves as genomes of progeny virions. Although the signal for switching from primary transcription to replication has not been conclusively identified for hantaviruses, a likely candidate is accumulation of N (translated from newly synthesized mRNA). The availability of N might drive its association with the vRNA template, perhaps changing the structure of the vRNA to increase processivity (i.e., allow RdRp synthesis through transcription termination signals). Although still speculative for hantaviruses, the importance of N for regulating transcription and replication of other negative-strand viruses, such as vesicular stomatitis virus (Patton et al. 1984) and influenza virus (Honda et al. 1988; Shapiro and Krug 1988; Li et al. 1998) is well documented.

3.2 Initiation of mRNA and vRNA/cRNA Synthesis

For hantaviruses, and other viruses in the family *Bunyaviridae*, transcription initiation requires the continuous production of suitable primers, which the RdRp acquires from capped cellular mRNAs. Similar to the mechanism described for

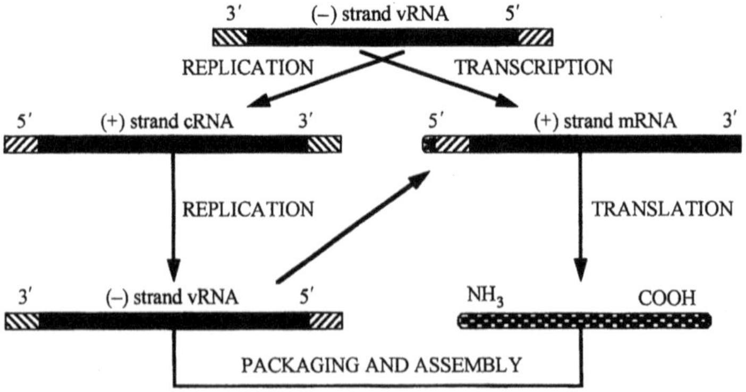

Fig. 4. Hantavirus replication and transcription pathway

influenza virus (Krug et al. 1981), the RdRp of hantaviruses (Garcin et al. 1995) and other viruses in the family (reviewed in Schmaljohn 1996a) catalyzes an endonucleolytic cleavage 7–18 nucleotides downstream of the 5′ cap of host cell mRNAs. However, in contrast to influenza virus, which obtains primers from newly synthesized mRNAs in the cell's nucleus, the RdRps of viruses in the family *Bunyaviridae* cleave primers from mRNAs in the cytoplasm (Rossier et al. 1986).

Sequence analysis of the 5′ ends of hantavirus mRNAs reveals the heterogeneity of the primer population and the occasional loss of one of the conserved triplet sequences (5′ UAG) (Garcin et al. 1995; Hutchinson et al. 1996; Meyer and Schmaljohn 2000a). Thus, transcription results in a viral mRNA population with 5′ extensions consisting of the host-derived primers. Despite the overall heterogeneity of the primers, hantaviral mRNAs have a preponderance of terminal G residues adjacent to the viral nucleotides. This finding is consistent with a model proposed for the initiation of hantaviral transcription and replication in which host-derived primers with terminal G residues are preferred (Garcin et al. 1995). The model, termed "prime-and-realign" suggests that the terminal G residue aligns with the third nucleotide of the hantaviral RNA template (C residue) to initiate RdRp activity. After synthesis of a few nucleotides, the nascent RNA realigns by slipping backward three nucleotides on the repeated terminal sequences (AUCAUCAUC) of the L, M, and S RNAs. The outcome of the slippage is that the G becomes the first nucleotide of the 5′ extensions. This realignment of the G to position −1 produces an exact copy of the 3′ end of the viral RNA. However, if the G initiates at position +6, rather than +3, and realigns to position +3, the first two nucleotides of the vRNA will be lost, a finding that is commonly observed in sequence studies. Whether this deletion has an effect on translation or message stability is unknown.

A similar prime-and-realign model for vRNA and cRNA synthesis was proposed, except that transcription was suggested to initiate with pppG alignment at the third nucleotide (C residue) of the template RNA (Garcin et al. 1995). As with mRNA synthesis, after the addition of a few nucleotides, polymerase slipping realigns the nascent RNA such that the initial priming G residue overhangs the template. In this situation, however, nucleolytic activity of the L protein might remove the overhanging G, leaving a monophosphorylated U residue at the nascent 5′ end. The presence of the monophosphorylated U on Hantaan virus RNA was experimentally demonstrated. Although data obtained for other viruses in the family (e.g., phleboviruses and tospoviruses) are consistent with the prime-and-realign model, it remains to be determined if this is a required or preferred mechanism of RdRp activity (Vialat and Bouloy 1992; Jin and Elliott 1993a,b; Prehaud et al. 1997; Duijsings et al. 1999).

3.3 Elongation and Transcription Termination

Elongation of the mRNA transcript by the RdRp requires the displacement of N from the viral templates. Although not shown experimentally, the RdRp may provide this activity. In addition, reloading of N onto the vRNA and unwinding of the newly synthesized mRNA from the vRNA must occur behind the transcription complex. The synthesis of the hantaviral mRNAs terminates shortly beyond the coding information, as indicated by nearly genomic length mRNAs in virus-infected cells (SCHMALJOHN et al. 1987) and by sequence analysis of mRNAs obtained by RT-PCR (HUTCHINSON et al. 1996; MEYER and SCHMALJOHN 2000a,b). This finding is consistent with data collected for other viruses in the family *Bunyaviridae*, which indicate that about 100 nucleotides found at the 3′ ends of cRNA are missing in mRNA (PATTNAIK and ABRAHAM 1983; BOULOY et al. 1984; PATTERSON et al. 1984; ESHITA et al. 1985; PETTERSSON et al. 1985; COLLETT 1986; RAJU and KOLAKOFSKY 1987; BOULOY et al. 1990). Potential U-rich sequence motifs are proposed as transcription termination sites for several viruses in the family, including the M segment of Sin Nombre virus (ROBERTSON et al. 1981; GUPTA and KINGSBURY 1982; ESHITA and BISHOP 1984; ESHITA et al. 1985; BOULOY et al. 1990; HUTCHINSON et al. 1996). Although such sites are similar to polyadenylation signals used by other negative-strand RNA viruses, there is little evidence that mRNAs of hantaviruses, or viruses in any of the other genera of the family, are polyadenylated (ULMANEN et al. 1981; PATTNAIK and ABRAHAM 1983; PETTERSSON et al. 1985). Moreover, transcription termination sites without the homopolymeric U5 or U6 tract are reported for several gene segments of viruses in the family, including the S segment of Sin Nombre virus (ESHITA and BISHOP 1984; COLLETT 1986; HUTCHINSON et al. 1996). It is possible that primary sequence alone is insufficient for signaling transcription termination, but instead, a secondary structure of the RNA is needed. Perhaps this structure only exists in naked RNAs, such as mRNAs, and is eliminated when it is complexed with N (as are vRNA and cRNA). In addition, other viral or cellular proteins may play a role in transcription termination. This idea is supported by a study with the expressed polymerase protein of Bunyamwera virus, which was infrequently able to correctly terminate transcription of mRNAs from virion-derived nucleocapsids. Currently, there is no consistent model for the process of mRNA transcription termination for hantaviruses and other viruses in the family *Bunyaviridae*.

4 Virion Assembly

The pathways of hantaviral assembly and release are reviewed elsewhere in this volume. Factors dictating selective encapsidation of each of the three unique gene segments into virions have not been fully determined. Because hantaviruses do not have a matrix protein to connect N and the envelope proteins, it is likely that there is a direct interaction between the cytoplasmic tails of G1 and/or G2 and N. The regions of these protein–protein interactions have not been defined for hantaviruses, but it is likely that this interaction directs the inclusion of ribonucleocapsids, and conversely the exclusion of mRNA, from virions. However, cRNA is usually also excluded from virions, although it is complexed with N. Thus, other factors, perhaps secondary or tertiary RNA structures formed by the genomic RNA, must contribute to assembly.

Although it is generally assumed that virions contain one each of the L, M, and S genome segments, few data exist to support this. Clearly, at least one of each segment is required to impart infectivity, but equimolar amounts may not be. In a study of segment reassortment between two Sin Nombre virus genotypes, diploid viruses (i.e., containing S or M segments from both genotypes) were identified within single, isolated plaques (RODRIGUEZ et al. 1998). Similarly, in cell cultures infected with two distinct bunyaviruses, La Crosse and snowshoe hare viruses, 9% of plaques recovered appeared to have corresponding segments from both viruses (usually S; URQUIDI and BISHOP 1992). In the same study, statistical analysis of the genome complexities indicated that segment reassortment was not random, but that there was a preference for homologous gene segments. What factors are responsible for limiting the number and type of gene segments included in virions are not yet defined.

5 Hantaviral Persistence

Hantaviruses cause persistent infections in cell cultures and in rodents. Surprisingly little is known about the biological characteristics of persistence, and even less is known about molecular properties that contribute to persistence. It is generally thought that once a natural rodent reservoir is infected with a hantavirus the animal remains infected for life and suffers no ill effects from the infection. This might be true, but data to support this are currently available for only four hantaviruses and their reservoirs: Hantaan virus (LEE et al. 1981), Seoul virus (LEE et al. 1982; TANISHITA et al. 1986; DOHMAE et al. 1994; KARIWA et al. 1996), Puumala virus (BOGDANOVA et al. 1987) and Black Creek Canal virus (HUTCHINSON et al. 1998). In these studies, both persistence in tissues and virus shedding were demonstrated, although conditions of the infections and methods used to detect virus varied considerably (reviewed in MEYER and SCHMALJOHN 2000b). Currently, it is not clear whether there are age-related factors permitting infection that leads to persistence, or if all ages permit establishment of persistence. In laboratory studies, infection of newborn mice or rats with Hantaan virus or Seoul virus can be lethal (ZHANG 1984; KIM and MCKEE 1985; MCKEE et al. 1985; TANISHITA et al. 1986; ZHANG et al. 1988; ZHANG et al. 1989; YOO et al. 1993). However, for the hantaviruses studied, vertical transmission does not occur (LEE et al. 1985; MORITA et al. 1993), and maternal antibody protects newborns from infection (ZHANG et al. 1988; DOHMAE and NISHIMUNE 1995). Therefore, in nature, newborns are probably rarely infected. Current evidence suggests that rodents with competent immune systems can be infected with hantaviruses, and despite the development of vigorous antibody responses to the infecting virus, those rodents develop long-term infections.

The ability of hantaviruses to survive in nature depends on their ability to maintain viral genetic information in rodents, while simultaneously avoiding the host immune system. How this is accomplished is not known. Both neutralizing antibodies and cell-mediated immune responses have been elicited in hantavirus-infected rodents (LEE et al. 1981; TANISHITA et al. 1986; ARIKAWA et al. 1994; KARIWA et al. 1996). These findings suggest that characteristics of the virus itself are likely to have a role in maintaining persistence. Few studies, however, have addressed this aspect of persistence. An investigation of cell cultures persistently infected with Seoul virus indicated that terminal deletions in the viral gene segments occurred cyclically during persistence (MEYER and SCHMALJOHN 2000a,b). A correlation between the extent of the deletions and viral replication was observed. A model was proposed suggesting that the deletions were a mechanism for downregulating viral replication to a level that would not be cytopathic. Deletions were also observed on the termini of the L and S segments of Tula virus examined 3–12 days after infection (KUKKONEN et al. 1998). In that study, the deleted termini were observed throughout the short-term infection; however, the infection was not observed long enough to determine if cyclical recovery of the terminal nucleotides occurred as it did in the Seoul virus experiments.

No data are currently available to support the terminal nucleotide deletion model as a mechanism of persistence in rodents, although downregulation of viral replication was clearly demonstrated for BCCV, as persistence progressed in infected cotton rats (HUTCHINSON et al. 1998). Obviously, much remains to be learned about the molecular and immunological factors involved in hantavirus persistence, and it is likely that both play crucial roles in the ability of hantaviruses to survive in nature.

6 Genetic Approaches to Transcription and Replication

Recent progress toward generating reverse genetics systems for negative-strand RNA viruses offers hope for the development of such systems for hantaviruses. To date, an infectious clone has been reported for only one virus in the family *Bunyaviridae*, Bunyamwera virus, and even that one has proven too difficult to use for routine genetic studies (BRIDGEN and ELLIOTT 1996). Problems in generating reverse genetic systems are ascribed to the unique replication pathways for these viruses, which require precise 5′ and 3′ ends of genomic and antigenomic templates, and the requirement for the RNA templates to be complexed with N. Nevertheless, several systems have been developed to analyze transcription and replication of negative-strand RNA viruses. These systems do not depend on production of infectious virus but instead employ expressed L and authentic viral templates. In the *Bunyaviridae* family, such systems have been reported for Bunyamwera virus, (JIN and ELLIOTT 1992; DUNN et al. 1995) and Rift Valley fever virus, (LOPEZ et al. 1995). These systems have proven useful for analyzing the basic properties of transcription and replication, e.g., that L and N are necessary and sufficient for in vitro transcription (LOPEZ et al. 1995), and for determining the importance of the terminal nucleotides for transcription (PREHAUD et al. 1997; ELLIOTT and DUNN 1999). Study of hantaviral replication would be greatly facilitated by the availability of an in vitro replication system or infectious clone. Such systems would provide researchers the ability to move genes easily in and out of the cell for mutational analysis of the individual genomic segments, and thereby allow biochemical and molecular contributions of the individual protein components to be understood.

References

Arikawa J, Ito M, Yao JS, Kariwa H, Takashima I, Hashimoto N (1994) Epizootiological studies of hantavirus infection among urban rats in Hokkaido, Japan: evidences for the persistent infection from the sero-epizootiological surveys and antigenic characterizations of hantavirus isolates. J Vet Med Sci 56:27–32

Betenbaugh M, Yu M, Kuehl K, White J, Pennock D, Spik K, Schmaljohn C (1995) Nucleocapsid- and virus-like particles assemble in cells infected with recombinant baculoviruses or vaccinia viruses expressing the M and the S segments of Hantaan virus. Virus Res 38:111–124

Bogdanova S, Gavrilovskaya I, Boyko V, Prokhorova N, Linev M, Apekina N, Gorbachkova Y, Rymalov N, Bernshteyn A, Chumakov M (1987) Persistent infection caused by hemorrhagic fever with renal syndrome virus in bank voles (Clethrionomys glareolus) – the natural host of the virus. Mikrobiologicheskiy Zhurnal 49:99–106

Bouloy M, Pardigon N, Vialat P, Gerbaud S, Girard M (1990) Characterization of the 5′ and 3′ ends of viral messenger RNAs Isolated from BHK21 cells infected with Germiston virus (Bunyavirus). Virology 175:50–58

Bouloy M, Vialat M, Girard M, Pardigon N (1984) A transcript from the S segment of the Germiston bunyavirus is uncapped and codes for the nucleoprotein and a nonstructural protein. J Virol 49: 717–723

Bridgen A, Elliott RM (1996) Rescue of a segmented negative-strand RNA virus entirely from cloned complementary DNAs. Proc Natl Acad Sci USA 93:15400–15404

Collett MS (1986) Messenger RNA of the M segment RNA of Rift Valley fever virus. Virology 151: 151–156

Dohmae K, Nishimune Y (1995) Protection against hantavirus infection by dam's immunity transferred vertically to neonates. Arch Virol 140:165–172

Dohmae K, Okabe M, Nishimune Y (1994) Experimental transmission of hantavirus infection in laboratory rats. J Infect Dis 170:1589–1592

Domingo E, Holland JJ (1994) Mutation rates and rapid evolution of viruses. In: Morse SS (ed) The Evolutionary Biology of Viruses. Raven Press, New York

Draper DE (1995) Protein-RNA recognition. Annu Rev Biochem 64:593–620

Duijsings D, Kormelink R, Goldbach R (1999) Alfalfa mosaic virus RNAs serve as cap donors for tomato spotted wilt virus transcription during coinfection of Nicotiana benthamiana. J Virol 73: 5172–5175

Dunn EF, Pritlove DC, Jin H, Elliott RM (1995) Transcription of a recombinant bunyavirus RNA template by transiently expressed bunyavirus proteins. Virology 211:133–143

Elliott R (1996) The *Bunyaviridae*: Concluding remarks and future prospects. In: Elliott RM (ed) The *Bunyaviridae*. Plenum Press, New York

Elliott RM, Dunn E (1999) Analysis of bunyavirus promoter sequences. XI International Congress of Virology, Sydney, Australia

Elton D, Metcalf L, Bishop K, Harrison D, Digard P (1999) Identification of amino acid residues of influenza virus nucleoprotein essential for RNA binding. J Virol 73:7357–7367

Eshita Y, Bishop DH (1984) The complete sequence of the M RNA of snowshoe hare bunyavirus reveals the presence of internal hydrophobic domains in the viral glycoprotein. Virology 137:227–240

Eshita Y, Ericson B, Romanowski V, Bishop DH (1985) Analyses of the mRNA transcription processes of snowshoe hare bunyavirus S and M RNA species. J Virol 55:681–689
Fodor E, Pritlove DC, Brownlee GG (1994) The influenza virus panhandle is involved in the initiation of transcription. J Virol 68:4092–4096
Garcin D, Lezzi M, Dobbs M, Elliott RM, Schmaljohn C, Kang CY, Kolakofsky D (1995) The 5′ ends of Hantaan virus (*Bunyaviridae*) RNAs suggest a prime-and-realign mechanism for the initiation of RNA synthesis. J Virol 69:5754–5762
Gott P, Stohwasser R, Schnitzler P, Darai G, Bautz EK (1993) RNA binding of recombinant nucleocapsid proteins of hantaviruses. Virology 194:332–337
Gupta KC, Kingsbury DW (1982) Conserved polyadenylation signals in two negative-strand RNA virus families. Virology 120:518–523
Hansen JL, Long AM, Schultz SC (1997) Structure of the RNA-dependent RNA polymerase of poliovirus. Structure 5:1109–1122
Henikoff S, Henikoff JG (1994) Protein family classification based on searching a database of blocks. Genomics 19:97–107
Hewlett MJ, Pettersson RF, Baltimore D (1977) Circular forms of Uukuniemi virion RNA: an electron microscopic study. J Virol 21:1085–1093
Hofmann K, Bucher P, Falquet L, Bairoch A (1999) The PROSITE database, its status in 1999. Nucl Acids Res 27:215–219
Honda A, Ueda K, Nagata K, Ishihama A (1988) RNA polymerase of influenza virus: role of NP in RNA chain elongation. J Biochem 104:1021–1026
Hughes SH, Stock AM (in press) Preparing recombinant proteins for X-ray crystallography. In: Rossmann MG, Arnold E (eds) International Tables for Macromolecular Crystallography, Volume F. Kluwer Academic Publisher, Dordrecht
Hutchinson KL, Peters CJ, Nichol ST (1996) Sin Nombre virus mRNA synthesis. Virology 224:139–149
Hutchinson KL, Rollin PE, Peters CJ (1998) Pathogenesis of a North American hantavirus, Black Creek Canal virus, in experimentally infected *Sigmodon hispidus*. Am J Trop Med Hyg 59:58–65
Jin H, Elliott RM (1992) Mutagenesis of the L protein encoded by Bunyamwera virus and production of monospecific antibodies. J Gen Virol 73:2235–2244
Jin H, Elliott RM (1993a) Characterization of Bunyamwera virus S RNA that is transcribed and replicated by the L protein expressed from recombinant vaccinia virus. J Virol 67:1396–1404
Jin H, Elliott RM (1993b) Non-viral sequences at the 5′ ends of Dugbe nairovirus S mRNAs. J Gen Virol 74:2293–2297
Kariwa H, Kimura M, Yoshizumi S, Arikawa J, Yoshimatsu K, Takashima I, Hashimoto N (1996) Modes of Seoul virus infections: persistency in newborn rats and transiency in adult rats. Arch Virol 141:2327–2338
Kim GR, McKee KT Jr (1985) Pathogenesis of Hantaan virus infection in suckling mice: clinical, virologic, and serologic observations. Am J Trop Med Hyg 34:388–395
Kolakofsky D, Hacker D (1991) Bunyavirus RNA synthesis: genome transcription and replication. Curr Top Microbiol Immunol 169:143–159
Krug RM, Plotch SJ, Ulmanen I, Herz C, Bouloy M (1981) The mechanism of initiation of influenza viral RNA transcription by capped RNA primers. In: Bishop DHL, Compans RW (eds) The Replication of Negative Strand Viruses. Elsevier North Holland, New York
Kukkonen SK, Vaheri A, Plyusnin A (1998) Completion of the Tula hantavirus genome sequence: properties of the L segment and heterogeneity found in the 3′ termini of S and L genome RNAs. J Gen Virol 79:2615–2622
Lee H, Baek L, Johnson K (1982) Isolation of Hantaan virus, the etiologic agent of Korean hemorrhagic fever from wild urban rats. J Infect Dis 146:638–644
Lee H, Baek L, Joo Y, Ahn I, Park K (1985) Study of horizontal and vertical transmission of Hantaan and Seoul virus in Apodemus agrarius and in rats. J Korean Soc Virol 15:55–63
Lee H, Lee P, Baek L, Song C, Seong I (1981) Intraspecific transmission of Hantaan virus, etiologic agent of Korean hemorrhagic fever, in the rodent *Apodemus agrarius*. Am Soc Trop Med Hyg 30:1106–1112
Li M, Ramirez BC, Krug RM (1998) RNA-dependent activation of primer RNA production by influenza virus polymerase: different regions of the same protein subunit constitute the two required RNA-binding site. EMBO J 17:5844–5852
Lopez N, Muller R, Prehaud C, Bouloy M (1995) The L protein of Rift Valley fever virus can rescue viral ribonucleoproteins and transcribe synthetic genome-like RNA molecules. J Virol 69:3972–3979
Luytes W, Krystal M, Enami M, Parvin J, Palese P (1989) Amplification, expression, and packaging of a foreign gene by influenza virus. Cell 59:1107–1113

Mathews DH, Sabina J, Zucker M, Turner DH (1999) Expanded sequence dependence of thermodynamic parameters provides robust prediction of RNA secondary structure. J Mol Biol 288:911–940
McKee K Jr, Kim G, Green D, Peters C (1985) Hantaan virus infection in suckling mice: virologic and pathologic correlates. J Med Virol 17:107–117
Meyer BJ, Schmaljohn C (2000a) Accumulation of terminally deleted RNAs may play a role in Seoul virus persistence. J Virol 74:1321–1331
Meyer BJ, Schmaljohn CS (2000b) Persistent hantavirus infections: characteristics and mechanisms. Trends Microbiol 8:61–67
Morita C, Morikawa S, Sugiyama K, Komatsu T, Ueno H, Kitamura T (1993) Inability of a strain of Seoul virus to transmit itself vertically in rats. Jpn J Med Sci Biol 46:215–219
Muller R, Poch O, Delarue M, Bishop DH, Bouloy M (1994) Rift Valley fever virus L segment: correction of the sequence and possible functional role of newly identified regions conserved in RNA-dependent polymerases. J Gen Virol 75:1345–1352
O'Reilly EK, Kao CC (1998) Analysis of RNA-dependent RNA polymerase structure and function as guided by known polymerase structures and computer predictions of secondary structure. Virology 252:287–303
Patterson JL, Holloway B, Kolakofsky D (1984) La Crosse virions contain a primer-stimulated RNA polymerase and a methylated cap-dependent endonuclease. J Virol 52:215–222
Pattnaik AK, Abraham G (1983) Identification of four complementary RNA species in Akabane virus-infected cells. J Virol 47:452–462
Patton JT, Davis NL, Wertz GW (1984) Role of vesicular stomatitis virus proteins in RNA replication. In: Bishop D, Compans R (eds) Nonsegmented negative strand viruses. Paramyxoviruses and rhabdoviruses. Academic Press, Orlando
Pettersson RF, Kuismanen E, Rauonnholm R, Ulmanen I (1985) mRNAs of Uukuniemi virus, a bunyavirus. In: Becker Y (ed) Viral messenger RNA transcription, processing, splicing, and molecular structure. Nijhoff Publishing, Boston
Plyusnin A, Vapalahti O, Vaheri A (1996) Hantaviruses: genome structure, expression and evolution. J Gen Virol 77:2677–2687
Poch O, Sauvaget I, Delarue M, Tordo N (1989) Identification of four conserved motifs among the RNA-dependent polymerase encoding elements. EMBO Journal 8:3867–3875
Prehaud C, Lopez N, Blok MJ, Obry V, Bouloy M (1997) Analysis of the 3′ terminal sequence recognized by the Rift Valley fever virus transcription complex in its ambisense S segment. Virology 227:189–197
Raju R, Kolakofsky D (1987) Translational requirement of La Crosse virus S-mRNA synthesis. J Virol 63:122–128
Raju R, Kolakofsky D (1989) The ends of La Crosse virus genome and antigenome RNAs within nucleocapsids are base paired. J Virol 63:122–128
Robertson JS, Schubert M, Lazzarini RA (1981) Polyadenylation sites for influenza virus mRNA. J Virol 38:157–163
Rodriguez LL, Owens JH, Peters CJ, Nichol ST (1998) Genetic reassortment among viruses causing hantavirus pulmonary syndrome. Virology 242:99–106
Rossier C, Patterson J, Kolakofsky D (1986) La Crosse virus small genome mRNA is made in the cytoplasm. J Virol 58:647–650
Sankar S, Porter AG (1992) Point mutations which drastically affect the polymerization activity of encephalomyocarditis virus RNA-dependent RNA polymerase correspond to the active site of *Escherichia coli* DNA polymerase I. J Biol Chem 267:10168–10176
Schmaljohn CS (1996a) *Bunyaviridae*: The viruses and their replication. In: Fields BN, Knipe DM, Howley PM (eds) Fields Virology. Lippencott-Raven, Philadelphia
Schmaljohn CS (1996b) Molecular Biology of Hantaviruses. In: Elliott RM (ed) The *Bunyaviridae*. Plenum Press, New York
Schmaljohn CS (1998) Hantaviruses (*Bunyaviridae*). In: Webster RG, Granoff A (eds) Encyclopedia of Virology. W.B. Saunders, London
Schmaljohn CS, Dalrymple JM (1983) Analysis of Hantaan virus RNA: evidence for a new genus of *Bunyviridae*. Virology 131:482–491
Schmaljohn CS, Hasty SE, Harrison SA, Dalrymple JM (1983) Characterization of Hantaan virions, the prototype virus of hemorrhagic fever with renal syndrome. J Infect Dis 148:1005–1012
Seong BL, Brownlee GG (1992) Nucleotides 9 to 11 of the influenza A virion RNA promoter are crucial for activity in vitro. J Gen Virol 73:3115–3124
Severson W, Partin L, Schmaljohn CS, Jonsson CB (1999) Characterization of the Hantaan nucleocapsid protein-ribonucleic acid interaction. J Biol Chem 274:33732–33739

Shapiro GI, Krug RM (1988) Influenza virus RNA replication in vitro: synthesis of viral template RNAs and virion RNAs in the absence of an added primer. J Virol 62:2285–2290

Siomi H, Dreyfuss G (1997) RNA-binding proteins as regulators of gene expression. Curr Opin Gen Dev 7:345–353

Steitz TA (1999) DNA polymerases: structural diversity and common mechanisms. J Biol Chem 274:17395–17398

Tanishita O, Takahashi Y, Okuno Y, Tamura M, Asada H, Dantas J Jr, Yamanouchi T, Domae K, Kurata T, Yamanishi K (1986) Persistent infection of rats with haemorrhagic fever with renal syndrome virus and their antibody responses. J Gen Virol 67:2819–2824

Uhrig JF, Soellick TR, Minke CJ, Philipp C, Kellmann JW, Schreier PH (1999) Homotypic interaction and multimerization of nucleocapsid protein of tomato spotted wilt tospovirus: identification and characterization of two interacting domains. Proc Natl Acad Sci USA 96:55–60

Ulmanen I, Seppala P, Pettersson RF (1981) In vitro translation of Uukuniemi virus-specific RNAs: identification of a nonstructural protein and a precursor to the membrane glycoproteins. J Virol 37:72–79

Urquidi V, Bishop DHL (1992) Non-random reassortment between the tripartite RNA genomes of La Crosse and snowshoe hare viruses. J Gen Virol 73:2255–2265

Vialat P, Bouloy M (1992) Germiston virus transcriptase requires active 40S ribosomal subunits and utilizes capped cellular RNAs. J Virol 66:685–693

Wower IK, Wower J, Zimmermann RA (1998) Ribosomal protein L27 participates in both 50S subunit assembly and the peptidyl transferase reactions. J Biol Chem 273:19847–19852

Yoo YC, Yoshimatsu K, Yoshida R, Tamura M, Azuma I, Arikawa J (1993) Comparison of virulence between Seoul virus strain SR-11 and Hantaan virus strain 76–118 of hantaviruses in newborn mice. Microbiol Immunol 37:557–562

Zhang F (1984) Seroepidemiological investigation on hemorrhagic fever with renal syndrome. III. Carrier state and familial aggregation. Chung Hua Liu Hsing Ping Hsueh Tsa Chih 5:5–8

Zhang X, Takashima I, Hashimoto N (1988) Role of maternal antibody in protection from hemorrhagic fever with renal syndrome virus infection in rats. Arch Virol 103:253–265

Zhang X, Takashima I, Mori F, Hashimoto N (1989) Comparison of virulence between two strains of *Rattus* serotype hemorrhagic fever with renal syndrome HFRS virus in newborn rats. Microbiol Immunol 33:195–206

Zucker M, Mathews DH, Turner DH (1999) Algorithms and thermodynamics for RNA secondary structure prediction: a practical guide. In: Barciszewski J, Clark BFC (eds) RNA Biochemistry and Biotechnology. Kluwer Academic Publishers

Hantavirus Maturation

C.F. SPIROPOULOU

1 Introduction

During the process of virus maturation, all the viral structural components accumulate at the cell surface in order to interact and assemble into mature virions. The majority of the negative stranded RNA viruses have been shown to assemble and bud from the cell surface (CADD et al. 1997; GAROFF et al. 1998). In contrast, viruses of the family *Bunyaviridae* are characterized by an unusual pattern of intracellular maturation where the bilipid envelope of the virus particles is derived from the membranes of the endoplasmic reticulum-Golgi compartment (MURPHY et al. 1973; PETTERSSON and MELIN 1996). Due to this characteristic feature, Golgi maturation has been used as one of the criteria for classification of viruses as members of the family *Bunyaviridae*.

Hantaviruses are members of the family *Bunyaviridae* (SCHMALJOHN and DALRYMPLE 1983; SCHMALJOHN et al. 1985). Despite being rodent-borne viruses rather than arthropod-borne like the members of the other genera, the hantaviruses

Special Pathogens Branch, Division for Viral and Rickettsial Diseases, Centers for Disease Control and Prevention, 1600 Clifton Rd NE Atlanta, GA 30333, USA nevertheless have the same genomic structure and site for virus maturation. The functional genome is composed of three circularized RNA segments encapsidated in the N protein, forming the nucleocapsid structure. The nucleocapsid is packaged into the virus particle at the bilayered envelope where the two virus surface glycoproteins G1 and G2 are embedded (SCHMALJOHN 1996a). Extensive work on hantavirus ultrastructure and maturation was done during the 1980s using electron microscopy (EM) mainly on Hantaan virus (HTN), the prototype of the genus (HUNG et al. 1985). Intracellular maturation of hantaviruses can be observed, but not as easily as with some other members of the family *Bunyaviridae*. It was only recently, with the discovery of the New World hantaviruses, that the intracellular maturation of the hantaviruses has been questioned. EM studies on Sin Nombre (SNV) and Black Creek Canal (BCCV) viruses have suggested that the maturation of these viruses did not occur intracellularly in the Golgi compartment. This conclusion was based on the observation that the majority of the virus particles were present extracellularly, with only rare virus-like particles observed inside the cell (GOLDSMITH et al. 1995; RAVKOV et al. 1997). However, EM pathogenesis studies have revealed intracellular hantavirus-like particles in the microvascular lung endothelium and interstitial macrophages of HPS patients (ZAKI et al. 1995).

It is still a matter of some debate whether or not hantaviruses acquire their envelope from the Golgi apparatus or plasma membrane, or whether maturation occurs at both Golgi and plasma membranes.

We will discuss our current understanding of the hantaviruses' assembly and maturation, addressing the steps individually.

2 Virus Particle Morphology

The hantavirus particle consists of the nucleocapsid arranged internally as circular coils and the surface glycoproteins, G1 and G2, appearing as fuzzy surface projections, approximately 7nm in length (GOLDSMITH et al. 1995). The virion surface projections are composed of oligomers of the two glycoproteins. However, we do not yet know whether they are heterodimers or multimers (ANTIC et al. 1992). In EM negative staining, hantavirus virions appear with an unusual square grid-like structure on the surface. The virus particles are generally spherical in shape, with quite broad size variation, ranging from 70 to 210nm (HUNG 1988; GOLDSMITH et al. 1995). The presence of pleomorphic-sized particles could be explained by the addition of more than one copy of a genomic segment in the virion particle (RODRIGUEZ et al. 1998). Elongated virus-like particles are also seen, both in New and Old World hantaviruses, budding into the cisternae of the Golgi apparatus (GOLDSMITH et al. 1995). These show the same distinctive cross-striations, but are narrower (approximately 50nm) than the mature virus particles.

Unusually elongated tubular projections have also been observed at the surface of infected Vero E6 cells (Fig. 1), most prominently with the New World hanta-

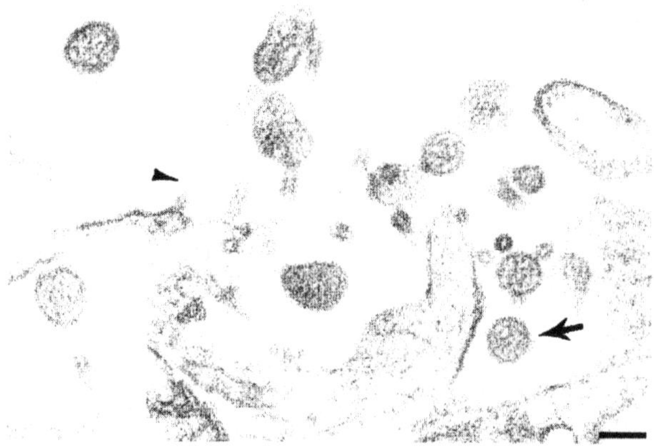

Fig. 1. Accumulation of extracellular SNV particles grown in Vero E6 cells (*arrow*). Numerous tubular projections are also present along the cell surface (*arrowheads*), and are specific for SNV glycoproteins as shown by immunogold labeling (not shown). Rare single virus particles are observed extracellularly in SNV infections of primary lung microvascular endothelial cells (*insert*). Bar: 100nm. (Courtesy of C.S. Goldsmith and L.H. Elliott; C.S. Goldsmith and C.F. Spiropoulou)

viruses (Goldsmith et al. 1995). Occasionally, these projections are also associated with virus particles. Structurally, they appear to be tubules 28nm wide and up to 550nm long. They can be immunostained for the virus glycoproteins but not the nucleocapsid. Their function, if any, is currently unknown.

3 Structural Organization of the Hantavirus Nucleocapsid Protein

In recent years, viral nucleocapsid proteins have been described as playing a protagonistic role in virus assembly and budding (Mebatsion et al. 1996; Garnier et al. 1998). In hantaviruses, nucleocapsid mRNA is the first viral RNA that accumulates during infection, with detectable amounts appearing as early as 6h post-infection (Hutchinson et al. 1996). Accordingly, virus nucleocapsid (N) protein also appears about the same time, as has been shown in immunofluorescence and immunoprecipitation studies (Schmaljohn 1996b). The N protein is the most abundant viral protein synthesized early in infections. In EM studies using immunogold labeling techniques, the nucleocapsid appears as granular and filamentous formations. These formations, called inclusion bodies, are very characteristic of hantavirus infections. The initial function assigned to the nucleocapsid is the protection of the virus genomic RNA from nuclease degradation, and participation in the formation of the ribonucleoprotein core (RNP) (Schmaljohn 1996b). The hantavirus nucleocapsid also plays an active role in the virus assembly. Unlike most of the other negative-stranded RNA viruses, the hantaviruses do not possess a matrix protein. It has been suggested that the nucleocapsid interacts directly with the cytoplasmic tail of the G1 for virus assembly to occur (Pettersson and Melin 1996). However, direct interaction of the nucleocapsid with either the G1 or the cytoplasmic tail of G1 has not yet been shown for any member of the family *Bunyaviridae*. The participation of the nucleocapsid in the virus assembly has also been implicated by an interesting study showing that the viral nucleocapsid associates with actin filaments during virus infections (RAVKOV and COMPANS 1998). The authors observed reduction of virus yields following treatment of the infected cells with cytochalasin D, an actin depolymerizing agent. They hypothesized that actin depolymerization influenced the virus production at the final steps of assembly and release, and that intact actin microfilaments are important in hantavirus morphogenesis. However, in other RNA viruses, cytochalasin D treatment was shown to reduce virus production at the stage of virus RNA replication, a step before virus assembly and release (GUPTA et al. 1998). This may be an alternative explanation for the observed reduction in virus titer. Involvement of actin in virus maturation has been shown for other viruses, such as alphaviruses and retroviruses (LAAKKONEN et al. 1998; LIU et al. 1999), and actin polymerization could also serve as an additional force for the membrane bending during budding (TSUKITA et al. 1997). Recently however, Ravkov and colleagues have reported the accumulation of expressed hantavirus nucleocapsid in the Golgi complex (RAVKOV and COMPANS 2000). The association of the nucleocapsid with the Golgi apparatus has also been shown previously in infections with Uukuniemi virus (UUK), another member of the family *Bunyaviridae*, which clearly matures in the Golgi (PETTERSSON and MELIN 1996; JANTTI et al. 1997). Expression of the Hantaan virus nucleocapsid protein alone, in insect or mammalian cells can lead to the formation of virus nucleocapsid-like structures without the assembly of virus-like particles (VLPs) (BETENBAUGH et al. 1995). Together, the above studies support the view that the nucleocapsid protein likely plays an important role in hantavirus assembly, but not as dominant a role as in the maturation of the retroviruses (GARNIER et al. 1998).

4 Synthesis and Trafficking of the Two Glycoproteins

4.1 Synthesis

The hantavirus M segment is relatively simple compared with that of some of the other members of the family *Bunyaviridae* in that only the G1 and G2 glycoproteins, but no NSm non-structural proteins, are encoded (SCHMALJOHN et al. 1987). The G1 and G2 proteins are generated by co-translational cleavage of the glycoprotein precursor (GPC), presumably by cellular signal peptidases present in the lumen of the endoplasmic reticulum (ER). No uncleaved GPC molecule has been detected in virus-infected cells or in cells expressing the M segment from a cDNA clone (SCHMALJOHN et al. 1987; PENSIERO et al. 1992). A schematic presentation of the GPC molecule and the G1 and G2 proteins is shown in Fig. 2. Based on direct sequencing of the amino terminus of the Hantaan virus G2 glycoprotein (SCHMALJOHN et al. 1987), it is presumed that the cleavage site for all hantaviruses is located immediately downstream of the conserved WAASA amino

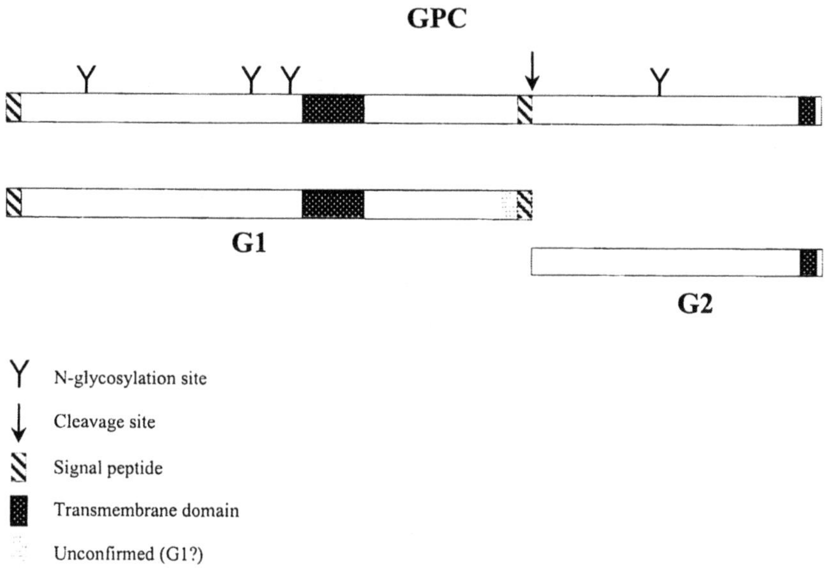

Fig. 2. Schematic diagram of the GPC precursor and G1/G2 glycoproteins

acid motif at the end of the G2 signal peptide. It is quite intriguing that although G1 and G2 are formed by cleavage from the GPC, they both possess their own functional signal sequences (SCHMALJOHN 1996b). The carboxy terminus of the mature G1 is not precisely known, but it appears to extend up to approximately 15 amino acids from the start of the G2 signal sequence (SCHMALJOHN et al. 1987; C.F. Spiropoulou et al., unpublished data). It is unknown if the G2 signal sequence is part of the mature G1, as has been shown for the G1 of UUK virus (ANDERSSON et al. 1997a). Interestingly, G2 can be expressed from a GPC clone even when the initiation codon for the G1 has been eliminated (KAMRUD and SCHMALJOHN 1994). This may be an artifact of the expression system used. However, the potential to synthesize G2 in the absence of G1 during infection may have some biological significance as G2 maturation appears to be the rate-limiting step in the correct folding of G1/G2 and their trafficking to the Golgi (see below).

4.2 Protein–Protein Interactions

In both Punta Toro and UUK viruses G1/G2 heterodimerization has been shown to be essential for their correct folding and transport from the ER to the Golgi apparatus (CHEN et al. 1991; PERSSON and PETTERSSON 1991). There are two findings suggesting protein–protein interactions between the hantavirus glycoproteins. Immunoprecipitation experiments, in which monoclonal antibodies either to G1 or G2 can co-precipitate both glycoproteins, even without prior cross-linking (ARIKAWA et al. 1989; ANTIC et al. 1992), is one. The other involves neutralizing epitope mapping studies. These also strongly suggest an interaction between G1 and G2, since most of the neutralizing epitopes are conformational epitopes, which are dependent on the presence of both proteins (WANG et al. 1993). In addition, the epitope mapping studies indicate that both G1 and G2 are present at the surface of the mature virions.

The G1 protein has an unusual tendency to aggregate (RUUSALA et al. 1992; C.F. Spiropoulou et al., unpublished data). The tendency of G1 to form self aggregates is prominent in sodium dodecyl sulfate (SDS) buffers during immunoprecipitation reactions in high temperatures and the presence of reducing agents. Similarly, aggregation has been seen for the equivalent of G1 protein of the Bunyamwera virus, in the *Bunyavirus* genus (NAKITARE and ELLIOTT 1993), the M protein of some of the coronaviruses (ROTTIER 1995), and the glycoproteins of hepatitis C virus (CHOUKHI et al. 1999).

The hantaviral G1 and G2 glycoproteins are both type I integral transmembrane proteins with C-terminal hydrophobic anchor domains. This means that their N-termini are facing towards the ER lumen and their C-termini are in the cytoplasm. G1 possesess an unusually long cytoplasmic tail (approx. 150 amino acids long), whereas G2 has a short, but highly charged cytoplasmic tail (less than 9 amino acids long). It has been speculated that the lengthy G1 cytoplasmic tail substitutes for the function of the matrix (M) protein which is present in many other RNA viruses. Even when viruses possess an M protein (e.g., rabies), the glycoprotein cytoplasmic tail seems to be involved in budding, as viruses with truncated forms of the glycoproteins bud relatively inefficiently (MEBATSION et al. 1996; SCHNELL et al. 1998). Also, by analogy with the model proposed for the maturation for alphaviruses, the hantavirus G1 tail would be expected to interact with the nucleocapsid (LOPEZ et al. 1994). The G1 tail could even interact directly with virion RNA, as has been seen with influenza virus (YE et al. 1999).

4.3 Glycosylation

Both G1 and G2 glycoproteins are modified by N-linked glycosylation. There are four potential N-glycosylation sites, which are conserved among all hantaviruses. Three of these are located in G1 (positions 142, 358, and 410 in SNV) and one in G2 (position 939; SPIROPOULOU et al. 1994). The G1 and G2 proteins acquire their sugars co-translationally in the ER. In general, their glycosylation status would be expected to change from high mannose to the complex type during their transport from the ER to the Golgi. Endo-H treatment, which cleaves high mannose-linked oligosaccharides, is frequently used to monitor the transport of the glycoproteins out of the ER to the Golgi complex. Analysis of transiently expressed SNV G1 and G2 glycoproteins by pulse-chase ^{35}S-Cys labeling and Endo-H treatment has shown that the proteins remain Endo-H sensitive even after a 6h chase (C.F. Spiropoulou et al., unpublished data). Identical results were seen with G1 and G2 proteins synthesized in SNV-infected Vero-E6 cells. Similar findings have been reported in earlier studies for Hantaan virus (SCHMALJOHN et al. 1986; ANTIC et al. 1992). The Endo-H sensitivity of the hantavirus glycoproteins is not restricted to only the glycoprotein present in infected cells, in that purified virion glycoproteins also remain Endo-H sensitive (ANTIC et al. 1992). Based on these results, it appears that the oligosaccharides linked to the hantavirus glycoproteins remain in the high mannose form, despite their apparent transport to the Golgi complex and possibly to the cell surface.

4.4 Glycoprotein Trafficking

As previously mentioned, one of the basic hallmarks of the members of the family *Bunyaviridae* is their unusual property of maturing at the Golgi apparatus (PETTERSSON and MELIN 1996). Golgi maturation is thought to be due to the retention and accumulation of the two viral glycoproteins, G1 and G2, in the Golgi complex (PETTERSSON and MELIN 1996). Based on this model, a number of studies have been done on members of the family *Bunyaviridae* to determine the characteristics of the two glycoproteins that contribute to their Golgi retention (MATSUOKA et al. 1988, 1992; PENSIERO and HAY 1992; RUUSALA et al. 1992; Nakitare and ELLIOTT 1993; ANDERSSON et al. 1997b). In general, it is thought that the glycoprotein closest to the N-terminus of the GPC precursor (equivalent to the G1 in the case of the hantaviruses) possesses the signal for Golgi localization (PETTERSSON and MELIN 1996). For example, the Golgi retention signal of Punta Toro virus has been mapped to the transmembrane domain and the first ten amino acids of the cytoplasmic tail of the G1 protein (MATSUOKA et al. 1994; MATSUOKA et al. 1996). For UUK virus, the Golgi retention signal has been mapped to the last 50 amino acids of the cytoplasmic tail of G1 (ANDERSSON et al. 1997b). These two examples illustrate that the Golgi retention signal is not sequence specific, since even Punta Toro and UUK viruses, which are both members of the *Phlebovirus* genus, use different amino acid sequences and even different domains of the G1 proteins. These findings are consistent with earlier studies, such as those with glycosyltransferases and coronavirus glycoproteins, which revealed that while retention in the Golgi apparatus is primarily mediated by the glycoprotein transmembrane domains, it does not require a precise amino acid sequence (MACHAMER and ROSE 1987; NILSSON et al. 1991; COLLEY et al. 1992; WONG et al. 1992). Taken together, these data suggest that knowledge of the retention signal for one of the bunyaviruses can not be generalized for the other members of the family *Bunyaviridae*.

A major limitation on the hantavirus maturation studies has been the difficulties in obtaining full-length clones for the GPC ORF from some hantaviruses. Panels of monoclonal antibodies needed for detection of different conformational stages during the glycoprotein maturation also are limited. Consequently, until recently studies have been limited to Hantaan virus, the prototype of the genus. However, there has been a recent surge of interest in the trafficking of hantavirus glycoproteins due to the discovery of HPS-associated hantaviruses (NICHOL et al. 1993) and the cell surface as the proposed site of virus maturation (GOLDSMITH et al. 1995; RAVKOV and COMPANS 1998).

If the intracellular localization of the glycoproteins is what defines the site of virus maturation in the family *Bunyaviridae*, it is of interest to determine whether the glycoproteins of the New or the Old World hantaviruses are retained in the Golgi complex or reach the cell surface before assembly and release. In addition, if the glycoproteins expressed from the GPC precursor are co-localizing in the Golgi complex, it is reasonable to assume that G1, G2 or both possess the Golgi retention signal. As far as the Old World hantaviruses are concerned, conflicting data exist concerning the individually expressed glycoproteins. In one study, the Hantaan virus G1 protein was reported to accumulate in the Golgi complex when expressed alone, while the G2 protein expressed alone was retained in the ER. (PENSIERO and HAY 1992). In another study, either Hantaan virus G1 or G2 expressed alone was reported to be retained in the ER (RUUSALA et al. 1992). However, both of these studies reported the Golgi localization of G1 and G2 proteins when the two proteins were co-expressed. In a more recent report, individually expressed G1 or G2 proteins of Hantaan virus were shown to exit the ER and reach the Golgi apparatus (ANHEIER et al. 1999). The domains of the two glycoproteins responsible for the exit from ER and their retention in the Golgi were identified in the cytoplasmic tail and the transmembrane domain, based on chimeric molecules and carboxy-end truncations. In our studies with Sin Nombre virus, we have found that both G1 and G2 glycoproteins expressed from the GPC precursor can exit the ER and become co-localized with the mannosidase II, a medial Golgi marker (Fig. 3D). There was no detection of the expressed glycoproteins at the cell surface. Interestingly, both G1 and G2, expressed individually, gave an immunofluorescence pattern quite different from that observed when they are expressed from the GPC precursor. Their exact localization was puzzling for some time. Partial co-localization with Golgi markers could be observed, but was dependent on the tissue culture cell lines used and the time course of the expression. However, the predominant immunofluorescence staining of the expressed proteins was observed in a defined area in close proximity to the Golgi complex (C.F. Spiropoulou et al., unpublished data). It appears the observed area represents the aggresome, a recently defined area where excess misfolded proteins accumulate (GARCIA-MATA et al. 1999).

While the fate of the individually expressed glycoproteins remains a topic for debate, there appears to be general agreement that virus glycoproteins of both the New and Old World hantaviruses, either expressed in infected cells, or from full-length GPC expressing clones, are retained in the Golgi (PENSIERO and HAY 1992; RUUSALA et al. 1992; C.F. Spiropoulou et al., unpublished data). This is also in accordance with the intracellular accumulation of the glycoproteins early on in the course of the hantavirus infection (Fig. 3A). It is only later in infection when the surface fluorescence labeling of the glycoproteins is observed (RAVKOV et al. 1997). This is no different from what is reported even with bunyaviruses, which are considered classic examples of viruses that mature at the Golgi (e.g., Uukuniemi virus), where surface labeling can also been seen in advanced stages of infection (PETTERSSON and MELIN 1996). In addition, while EM studies showing the presence of a "dense fuzzy layer" on the surface of Hantaan virus-infected cells had

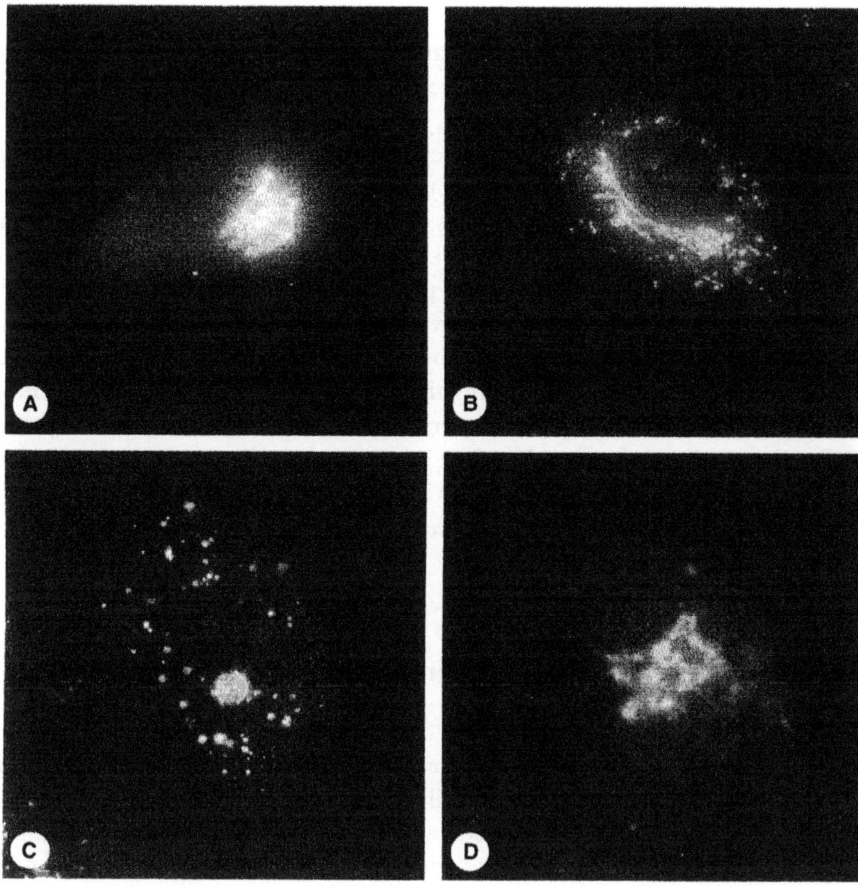

Fig. 3A,B. Time course of the SNV infection of primary human lung microvascular endothelial cells (HMVEC). Immunofluorescence staining using rabbit anti-SNV serum at day 1 (*A*) and day 5 (*B*). The immunofluorescence viral staining at day 1 was shown to be exactly co-localized with P58, a cis-Golgi marker (not shown). The co-localization was only partial by day 5, where more dispersed staining is seen. **C** Immunofluorescence on SNV-infected HMVEC cells stained with G1 monospecific antibody at day 5 post-infection shows the presence of G1 aggregates. **D** Localization of the expressed glycoproteins from the GPC clone. NRK cells were infected with vTF7-3 and transfected with a GPC-expressing clone. At 12h post-transfection, cycloheximide was added to the medium for 5h and then the cells were stained using rabbit anti-SNV serum. In double-stained experiments the glycoproteins where co-localized with mannosidase II, a medial Golgi marker (not shown)

earlier been interpreted as evidence of possible accumulation of virus antigens at the cell surface (HUNG 1988), later studies using immunogold-labeled hantavirus-specific antibodies showed no virus antigens associated with such layers (GOLDSMITH et al. 1995). Given the complex and sometimes contradictory data regarding a possible hantavirus Golgi retention signal, it remains unclear what the exact mechanism of Golgi retention may be. One possible mechanism of Golgi retention is that within the Golgi complex microenvironment G1 may form oligomers or lattice structures that are incompatible with transport (Weisz et al. 1993; Nilsson et al. 1994).

Although the Golgi retention signal appears to be independent of precise primary amino acid motifs, there are conserved sequence-specific signals at the end of the G1 and G2 cytoplasmic tails, which may influence the trafficking and localization of the glycoproteins. The cytoplasmic tail of the G1 glycoprotein contains a tyrosine-based motif (YXXL). This motif has been implicated in the internalization of surface glycoproteins, intracellular signaling and cell-to-cell spread of a number of viruses (Ohno et al. 1995; Tirabassi et al. 1998; Delamarre et al. 1999). The cytoplasmic tail of G2 possesses an ER-retention signal KKXX (Jackson et al. 1993). It had been suggested in earlier studies that this signal is functional when G2 is expressed alone, but during the interaction of G2 with G1, this signal is somehow hidden, allowing the G2 to exit the ER and enter the Golgi complex (Pensiero and Hay 1992). Recent studies have shown that such a signal is not functional when located within the 13 amino acids adjacent to the transmembrane domain (Vincent et al. 1998). This is likely the case in the hantaviruses, since the KKXX signal in G2 is directly adjacent to the transmembrane domain.

5 Possible Model of the Hantavirus Assembly and Release

With the recent surge in interest in hantaviruses, the assembly and maturation of the hantaviruses has been studied in detail. However, the accumulation of new data has challenged some of the early models and shown the overall picture to be much more complicated than originally thought, making it difficult to draw firm conclusions concerning the hantavirus assembly and maturation processes. There is a study directly addressing the involvement of the nucleocapsid and the two glycoproteins in Hantaan virus assembly (Betenbaugh et al. 1995). Its main purpose was to reconstitute virus-like particles (VLPs) from the virus' structural components and to use them in vaccine studies. The authors tried to produce VLPs using baculovirus or vaccinia virus recombinants expressing the N, G1, and G2 proteins in insect or mammalian cell lines, respectively. The outcome was that either the two glycoproteins or the nucleocapsid were unable to support budding on their own. Only when all three proteins were co-expressed in mammalian cells were they able to form VLPs. Unfortunately, the vaccinia recombinants used as the expression system are highly cytopathic, precluding studies of intracellular maturation by EM. In any case, the above results indicate that virus assembly is dependent on the nucleocapsid and both glycoproteins.

From the trafficking studies, it has been also shown that, in both the New and Old World hantaviruses, the expressed nucleocapsid and glycoproteins are accumulating intracellularly in the Golgi apparatus and that they could not be detected at the cell surface. The above information would support a model for intracellular maturation of the hantaviruses. However, such a picture would be incomplete. We also know that in advanced stages of virus infection, there is undeniable detection by immunofluorescence staining of substantial amounts of glycoproteins at the cell surface. Possible explanations for the presence of the two glycoproteins at the cell surface include: (a) The model for the New World hantavirus cell surface maturation is correct, in that actin interacts and transports the virus nucleocapsids to the plasma membrane, where the final step of virus assembly and release takes place (RAVKOV and COMPANS 1998). (b) Excessive accumulation of the G1 and G2 in the Golgi leads to saturation and leakage to the cell surface. (c) The glycoproteins recycle from the cell surface to the Golgi due to the presence of a recycling signal on the cytoplasmic tail of G1. (d) Already-released mature virions can reassociate with the cell surface. Taking all the above in consideration, it is possible that there are dual sites for hantavirus assembly. However, more studies are clearly needed before definitive conclusions can be drawn.

As for the site of virus release, studies have been limited to one member of the hantavirus genus. It has been shown that BCCV is released apically during infection of polarized Vero kidney epithelial cells (RAVKOV et al. 1997). Considering that hantavirus infection of humans is believed to be through aerosol transmission, apical release could make sense in terms of facilitating cell-to-cell spread of the virus in the respiratory tract. However, it is difficult to understand how the systemic part of the infection is initiated in this manner. Macrophages or dendritic cell populations on the epithelium may be the Trojan horses allowing the virus to reach its major target, the endothelium.

6 Concluding Remarks

During the last decade, many new hantavirus sequences have been accumulated in the databases. New hantaviruses are continually being identified, and their phylogenetic relationship characterized in detail. However, we are only beginning to understand some of the cellular mechanisms by which the hantaviruses mature into infectious particles. Basic questions concerning virus maturation remain open. Does virus maturation occur intracellularly or at the cell surface? Do the Old World hantaviruses have a different mode of assembly than the New World viruses? What factors determine the site of virus maturation? Which one of the three major structural proteins is the driving force for virus maturation? Considering how complicated the field of virus maturation is in general, one can appreciate the amount of work still needed to answer these important questions. One of the desired approaches to tackle the virus maturation questions will be the future establishment of a reverse genetic system for hantaviruses similar to those developed for several other negative-stranded RNA viruses (SCHNELL et al. 1994; GARCIN et al. 1995; DURBIN et al. 1997; NEUMANN et al. 1999; PEKOSZ et al. 1999). Unfortunately, reverse genetics approaches have failed so far for members of the family *Bunyaviridae*, except for one member of the genus *Bunyavirus* (BRIDGEN and

Elliott 1996). Accumulating knowledge on virus maturation and assembly is very important in order to understand basic aspects of the hantavirus life cycle and pathogenesis. These studies could also give us insights for blocking different steps of virus maturation, and possibly giving rise to treatments for hantavirus diseases.

References

Andersson AM, Melin L, Persson R, Raschperger E, Wikstrom L, Pettersson RF (1997a) Processing and membrane topology of the spike proteins G1 and G2 of Uukuniemi virus. J Virol 71:218–225

Andersson AM, Melin L, Bean A, Pettersson RF (1997b) A retention signal necessary and sufficient for Golgi localization maps to the cytoplasmic tail of a *Bunyaviridae* (Uukuniemi virus) membrane glycoprotein. J Virol 71:4717–4727

Anheier B, Lober C, Lindow S, Klenk H-D, Feldmann H (1999) Intracellular maturation of Hantaan virus glycoproteins. Abstract, XIth International Congress of Virology (VP11.03)

Antic D, Wright KE, Kang CY (1992) Maturation of Hantaan virus glycoproteins G1 and G2. Virology 189:324–328

Arikawa J, Schmaljohn AL, Dalrymple JM, Schmaljohn CS (1989) Characterization of Hantaan virus envelope glycoprotein antigenic determinants defined by monoclonal antibodies. J Gen Virol 70:615–624

Betenbaugh M, Yu M, Kuehl K, White J, Pennock D, Spik K, Schmaljohn CS (1995) Nucleocapsid- and virus-like particles assemble in cells infected with recombinant baculoviruses or vaccinia viruses expressing the M and S segments of Hantaan virus. Virus Res 38:111–124

Bridgen A, Elliott RM (1996) Rescue of a segmented negative-strand RNA virus entirely from cloned complementary DNAs. Proc Natl Acad Sci USA 93:15400–15405

Cadd TL, Skoging U, Liljestrom P (1997) Budding of enveloped viruses from the plasma membrane. BioEssays 19:993–1000

Chen S-Y, Matsuoka Y, Compans RW (1992) Assembly of G1 and G2 glycoprotein oligomers in Punta Toro virus-infected cells. Virus Res 22:215–225

Choukhi A, Pillez A, Drobecq H, Sergheraert C, Wychowski C, Dubuisson J (1999) Characterization of aggregates of hepatitis C virus glycoproteins. J Gen Virol 80:3099–3107

Colley KJ, Lee EU, Paulson JC (1992) The signal anchor and stem regions of the beta-galactoside alpha 2,6-sialyltransferase may each act to localize the enzyme to the Golgi apparatus. J Biol Chem 267:7784–7793

Delamarre L, Pique C, Rosenberg AR, Blot V, Grange M-P, Le Blanc I, Dokhelar M-C (1999) The Y-S-L-I Tyrosine-based motif in the cytoplasmic domain of the human T-cell leukemia virus type 1 envelope is essential for cell-to-cell transmission. J Virol 73:9659–9663

Durbin AP, Hall SL, Siew JW, Whitehead SS, Collins PL, Murphy BR (1997) Recovery of infectious human parainfluenza virus type 3 from cDNA clone. Virology 235:323–332

Garcia-Mata R, Bebok Z, Sorscher EJ, Sztul ES (1999) Characterization and dynamics of aggresome formation by a cytosolic GFP-chimera. J Cell Biol 146:1239–1254

Garcin D, Pelet T, Calain PLR, Curran J, Kolakofsky D (1995) A highly recombinogenic system for the recovery of infectious Sendai paramyxovirus from cDNA: generation of a novel copy-back non-defective interfering virus. EMBO J 14:6087–6094

Garnier L, Bowzard B, Wills JW (1998) Recent advances and remaining problems in HIV assembly. AIDS 12:S5–S16

Garoff H, Hewson R, Opstelten D-JE (1998) Virus maturation by budding. Micr Mol Bio Rev 62:1171–1190

Goldsmith CS, Elliott LH, Peters CJ, Zaki SR (1995) Ultrastructural characteristics of Sin Nombre virus, causative agent of hantavirus pulmonary syndrome. Arch Virol 140:2107–2122

Gupta S, De BP, Drazba JA, Banergee AK (1998) Involvement of actin microfilaments in the replication of human parainfluenza virus type 3. J Virol 72:2655–2662

Hung TH, Zinyi C, Tungxin Z, Semao X, Changshou H (1985) Morphology and morphogenesis of viruses of hemorrhagic fever with renal syndrome (HFRS): I. Some peculiar aspects of the morphogenesis of various strains of HFRS virus. Intervirology 23:97–108

Hung TH (1988) Atlas of hemorrhagic fever with renal syndrome. Science Press, Beijing
Hutchinson KL, Peters CJ, Nichol ST (1996) Sin Nombre virus mRNA synthesis. Virology 224:139–149
Jackson MR, Nilsson T, Peterson PA (1993) Retrieval of transmembrane proteins to the endoplasmic reticulum. J Cell Biol 121:317–333
Jantti J, Hilden P, Ronka H, Makiranta V, Keranen S, Kuismanen E (1997) Immunocytochemical analysis of Uukuniemi virus budding compartments: role of the intermediate compartment and the Golgi stack in virus maturation. J Virol 71:1162–1172
Kamrud KI, Schmaljohn CS (1994) Expression strategy of the M genome segment of Hantaan virus. Virus Res 31:109–121
Laakkonen P, Auvinen P, Kujala P, Kaariainen L (1998) Alphavirus replicase protein NSP1 induces filopodia and rearrangement of actin filaments. J Virol 72:10256–10269
Liu B, Dai R, Tian C-J, Dawson L, Gorelick R, Yu X-F (1999) Interaction of the Human Immunodeficiency virus type 1 nucleocapsid with actin. J Virol 73:2901–2908
Lopez S, Yao J-S, Kuhn RJ, Strauss EG, Strauss JH (1994) Nucleocapsid-glycoprotein interactions required for assembly of alphaviruses. J Virol 68:1316–1323
Machamer C, Rose JK (1987) A specific transmembrane domain of a coronavirus E1 glycoprotein is required for its retention in the Golgi region. J Cell Biol 105:1205–1214
Matsuoka Y, Ihara T, Bishop DHL, Compans RW (1988) Intracellular accumulation of Punta Toro virus glycoproteins expressed from cloned cDNA. Virology 167:251–260
Matsuoka Y, Chen S-Y, Compans RW (1994) A signal for Golgi retention in the bunyavirus G1 glycoprotein. J Biol Chem 269:22565–22573
Matsuoka Y, Chen S-Y, Holland CE, Compans RW (1996) Molecular determinants of Golgi retention in the Punta Toro virus G1 protein. Arch Biochem Biophys 336:184–189
Mebatsion T, Konig M, Conzelmann K-K (1996) Budding of rabies virus particles in the absence of the spike glycoprotein. Cell 84:941–951
Murphy FA, Harrison AK, Whitfield SG (1973) *Bunyaviridae*: morphologic and morphogenetic similarities of Bunyamwera serologic supergroup viruses and several other arthropod-borne viruses. Intervirology 1:297–316
Nakitare GW, Elliott RM (1993) Expression of the Bunyamwera virus M genome segment and intracellular localization of NSM. Virology 195:511–520
Neumann G, Watanabe T, Ito H, Watanabe S, Goto H, Gao P, Hughes M, Perez DR, Donis R, Hoffmann E, Hobom G, Kawaoka Y (1999) Generation of influenza A viruses entirely from cloned cDNAs. Proc Natl Acad Sci USA 96:9345–9350
Nichol ST, Spiropoulou CF, Morzunov S, Rollin PE, Ksiazek TG, Feldmann H, Sanchez A, Childs J, Zaki S, Peters CJ (1993) Genetic identification of a hantavirus associated with an outbreak of acute respiratory illness. Science 262:914–917
Nilsson T, Lucocq JM, Mackay D, Warren G (1991) The membrane spanning domain of b-1,4-galactosyltransferase specifies trans Golgi localization. EMBO J 10:3567–3575
Nilsson T, Warren G (1994) Retention and retrieval in the endoplasmic reticulum and the Golgi apparatus. Curr Opin Cell Biol 6:517–521
Ohno H, Stewart J, Fournier M-C, Bosshart H, Rhee I, Miyatake S, Saito T, Gallusser A, Kirchhausen T, Bonifacino JS (1995) Interaction of tyrosine-based sorting signals with clathrin associated proteins. Science 269:1872–1874
Pekosz A, He B, Lamb RA (1999) Reverse genetics of negative-strand RNA viruses: closing the circle. Proc Natl Acad Sci USA 96:8804–8806
Pensiero MN, Hay J (1992) The Hantaan virus M-segment glycoproteins G1 and G2 can be expressed independently. J Virol 66:1907–1914
Persson R, Pettersson RF (1991) Formation and intracellular transport of a heterodimeric viral spike protein complex. J Cell Biol 112:257–266
Pettersson RF, Melin L (1996) Synthesis, assembly and intracellular transport of *Bunyaviridae* membrane proteins. In: Elliott RM (ed) The *Bunyaviridae*. Plenum Press, New York, pp 159–188
Ravkov EV, Nichol ST, Compans RW (1997) Polarized entry and release in epithelial cells of Black Creek Canal Virus, a New World hantavirus. J Virol 71:1147–1154
Ravkov EV, Compans RW (1998) Role of actin microfilaments in Black Creek Canal virus morphogenesis. J Virol 72:2865–2870
Ravkov EV, Compans RW (2000) Association of a hantavirus nucleocapsid protein with the Golgi complex. Abstract: cell biology of virus entry, replication and pathogenesis. Keystone symposia (216)
Rodriguez LL, Owens JH, Peters CJ, Nichol ST (1998) Genetic reassortment among viruses causing hantavirus pulmonary syndrome. Virology 242:99–106

Rottier PJM (1995) The Coronaviruses membrane glycoprotein. In: Siddell SG (ed) The Coronaviridae. Plenum Press, New York and London, pp 73–113

Ruusala A, Persson R, Schmaljohn CS, Pettersson RF (1992) Coexpression of the membrane glycoproteins G1 and G2 of Hantaan virus is required for targeting to the Golgi complex. Virology 186:53–64

Schmaljohn CS (1996a) Molecular biology of hantaviruses. In: Elliott RM (ed) The *Bunyaviridae*. Plenum Press, New York, pp 63–90

Schmaljohn CS (1996b) *Bunyaviridae* and their replication. In: Fields BN, Knipe DM, Howley PM (eds) Virology, Lippincott-Raven Pub, Philadelphia, pp 1447–1471

Schmaljohn CS, Dalrymple JM (1983) Analysis of Hantaan virus RNA: Evidence for a new genus of *Bunyaviridae*. Virology 131:482–491

Schmaljohn CS, Hasty SE, Dalrymple JM, LeDuc JW, Lee HW, von Bonsdorff C-H, Brummer-Korvenkontio M, Vaheri A, Tsai TF, Regnery HL, Goldgaber D, Lee P-W (1985) Antigenic and genetic properties of viruses linked to hemorrhagic fever with renal syndrome. Science 227:1041–1044

Schmaljohn CS, Hasty SE, Rasmussen L, Dalrymple JM (1986) Hantaan virus replication: effects of monensin, tunicamycin and endoglycosidases on the structural glycoproteins. J Gen Virol 67:707–717

Schmaljohn CS, Schmaljohn AL, Dalrymple JM (1987) Hantaan virus M RNA: coding strategy, nucleotide sequence, and gene order. Virology 157:31–39

Schnell MJ, Mebatsion T, Conzelmann K-K (1994) Infectious rabies viruses from cloned cDNA. EMBO J 13:4195–4203

Schnell MJ, Buonocore L, Boritz E, Ghosh HP, Chernish R, Rose JK (1998) Requirement of a non-specific glycoprotein cytoplasmic domain sequence to drive efficient budding of vesicular stomatitis virus. EMBO J 17:1289–1296

Spiropoulou CF, Morzunov S, Feldmann H, Sanchez A, Peters CJ, Nichol ST (1994) Genome structure and variability of a virus causing hantavirus pulmonary syndrome. Virology 200:715–723

Tirabassi RS, Enquist LW (1998) Role of envelope protein gE endocytosis in the pseudorabies virus life cycle. J Virol 72:4571–4579

Tsukita S, Yonemura S, Tsukita S (1997) ERM proteins: head to tail regulation of actin-plasma membrane interaction. Trends Biochem Sci 22:53–58

Vincent M, Martin A, Compans RW (1998) Function of the KKXX motif in endoplasmic reticulum retrieval of a transmembrane protein depends on the length and structure of the cytoplasmic domain. J Biol Chem 273:950–956

Wang M, Pennock DG, Spik KW, Schmaljohn CS (1993) Epitope mapping studies with neutralizing and non-neutralizing monoclonal antibodies to the G1 and G2 envelope glycoproteins of Hantaan virus. Virology 197:757–766

Weisz OA, Swift AM, Machamer CE (1993) Oligomerization of a membrane protein correlates with its retention in the Golgi complex. J Cell Biol 122:1185–1196

Wong SH, Low SH, Hong W (1992) The 17-residue transmembrane domain of β-galactoside α2,6-sialyltransferase is sufficient for Golgi retention. J Cell Biol 117:245–258

Ye Z, Liu T, Offringa DP, McInnis J, Levandowski RA (1999) Association of influenza virus matrix protein with ribonucleoproteins. J Virol 73:7467–7473

Zaki SR, Greer PW, Coffield LM, Goldsmith CS, Notle KB, Foucar K, Feddersen RM, Zumwalt RE, Miller GL, Khan AS, Rollin PE, Ksiazek TG, Nichol ST, Mahy BWJ, Peters CJ (1995) Hantavirus pulmonary syndrome: pathogenesis of an emerging infectious disease. Am J Pathol 146:552–579

Virus Evolution and Genetic Diversity of Hantaviruses and Their Rodent Hosts

A. PLYUSNIN*,** and S.P. MORZUNOV***

1 Introduction

Unlike other members of the *Bunyaviridae* family, which must be regarded as arboviruses[1], hantaviruses are not transmitted by arthropod vectors, and are

* Haartman Institute, Department of Virology, POB 21, University of Helsinki, 00014 Helsinki, Finland
** Karolinska Institute, 17177 Stockholm, Sweden
*** Department of Microbiology, MS 320, University of Nevada-Reno, Reno, Nevada 89557, USA

[1] In fact, members of the family *Bunyaviridae* may display alternative survival strategies, depending on the nature of their specific host organisms, e.g., the plant-infecting tospoviruses are not true arboviruses either.

exclusively maintained in the populations of their specific rodent hosts. Thus, the prefix "Robo" (from ROdent-BOrne) seems to be more appropriate for the viruses constituting the genus *Hantavirus*[2]. Each of the over 20 currently recognized hantavirus species is predominantly associated with one (or a few closely related) specific rodent species in which it establishes persistent infection. The fact that rodents, or more specifically, rodents alone, constitute the entire host range within which hantaviruses evolve, has several important consequences: (a) Distinctive characteristics of different hantaviruses are formed as adaptations to the distinct genetic environment of their rodent hosts; (b) Contemporary distribution of distinct hantaviruses reflects complicated history of co-speciation events and rodents' migrations (the most recent extensive rodent migrations were caused by the sequence of several glaciation and deglaciation events in the northern hemisphere). This forms a basis for the circulation of distinct hantaviruses on different continents, their co-existence in some geographic regions, and geographical clustering of hantavirus genetic variants; (c) As a general rule, humans are merely evolutionary "dead-end" hosts for hantaviruses, and thus, human epidemics do not contribute to virus evolutionary process[3]. As humans are usually naive towards hantaviruses as antigens, that results in sometime dramatic immunological "excesses" leading to high mortality of known hantavirus-caused diseases, severe hemorrhagic fever with renal syndrome (HFRS) (up to 10–12%) and hantavirus pulmonary syndrome (HPS) (up to 40–50%). Thus, reasonable laboratory models to study hantavirus infections and associated diseases should include primates.

In this review we intend to show that hantavirus phylogeny, as well as their contemporary genetic diversity and geographic distribution, mirrors those of the natural hosts as a result of long-term co-evolution through co-speciation, sorting, and host-switching events.

2 Old World Hantaviruses and Their Rodent Hosts

2.1 Murinae-Borne Hantaviruses

Phylogenetically, hantaviruses carried by Murinae rodents (the Old World mice and rats) form a well-supported group separated from the two other groups which comprise hantaviruses harbored by Arvicolinae rodents (voles and lemmings, which

[2] For the sake of simplicity, throughout this review we will be discussing rodents as the natural hosts for hantaviruses without attempting to resolve the exceptional case of the Thottapalayam virus, which was isolated from a shrew, *Suncus murinus*, in the 1970s (Carey et al. 1971).

[3] Considering the extremely high mutation rate and adaptability of RNA viruses, we cannot exclude a possibility of emergence of novel hantavirus strains with altered pathogenicity and increased transmissibility as a result of a human outbreak of a hantavirus disease in the future. As for now, at least one hantavirus has been convincingly proven to be capable for "human-to-human" transmission (Padula et al. 1998).

are distributed throughout the entire northern hemisphere) and Sigmodontinae rodents (the New World mice and rats) (Fig. 1).

2.1.1 Hantaan Virus

Hantaan (HTN) virus, a prototype virus of the *Hantavirus* genus, was isolated from the lung tissue of the striped field mouse, *Apodemus agrarius*, trapped near the

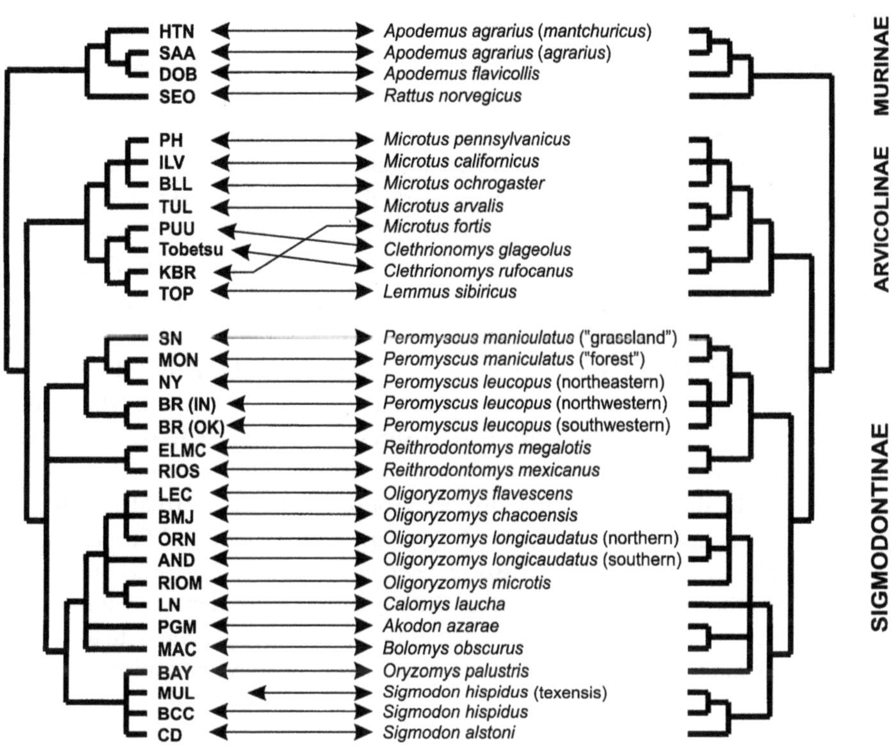

Fig. 1. Congruence of rodent and hantavirus phylogenies. Facing each other cladograms summarize our current knowledge of the phylogenies of hantaviruses and their rodent hosts. Virus phylogeny is based on the recent phylogenetic reconstructions published by BHARADWAJ et al. 1997; FULHORST et al. 1997; JOHNSON et al. 1997; LEVIS et al. 1998; MORZUNOV et al. 1998; NICHOL et al. 1999; and VAPALAHTI et al. 1999. Rodent phylogeny is based primarily on the reconstructions published by DICKERMAN and YATES (1995); ENGEL et al. (1998); SMITH and PATTON (1993). Virus data were derived mainly from the M genome segment-based trees. However, S-segment phylogenetic trees were also used where M-segment data for corresponding viruses were not available. In both hantavirus and rodent cladograms only branches that had high bootstrap support (over 70%) and were consistently present in all phylogenetic trees analyzed are shown. Branches with low bootstrap support, and clearly "unresolved" branches (i.e., branches that were placed differently in the different phylogenetic trees analyzed) are drawn as ones coming from the same resolved ancestral node. Viruses: *HTN* Hantaan, *SAA* Saaremaa, *DOB* Dobrava, *SEO* Seoul, *PH* Prospect Hill, *ILV* Isla Vista, *BLL* Bloodland Lake, *TUL* Tula, *PUU* Puumala, *KBR* Khabarovsk, *TOP* Topografov, *SN* Sin Nombre, *MON* Monongahela, *NY* New York, *BR (IN)* Blue River (Indiana lineage), *BR (OK)* Blue River (Oklahoma lineage), *ELMC* El Moro Canyon, *RIOS* Rio Segundo, *LEC* Lechiguanas, *BMJ* Bermejo, *ORN* Oran, *AND* Andes, *RIOM* Rio Mamore, *LN* Laguna Negra, *PGM* Pergamino, *MAC* Maciel, *BAY* Bayou, *BCC* Black Creek Canal, *CD* Caño Delgadito

Hantaan river in South Korea (Lee and Lee 1978). HTN is endemic in the Far East causing HFRS cases with mortality up to 10% (Lee et al. 1990).

Complete M-segment sequences determined for the prototype *A. agrarius* HTN virus strain 76–118 showed diversity of 1.6–5.7% with two isolates from Korean HFRS patients (Schmaljohn et al. 1988) and 15.3% with the HTN virus isolate HV114 originating from a Chinese HFRS patient (Xiao et al. 1993a). Partial M-segment sequences of nine Chinese isolates (two of mice origin, and seven of human origin) fell into three groups which diverged by 17.1–21.9% from each other and by 14.1–22.5% from the strain 76–118 (Liang et al. 1994). While the Korean strains form a distinct, well-supported phylogenetic lineage, the Chinese strains show more heterogeneity: HV114 and A9 strains are clustered together and form a separate lineage; a separate position of another Chinese strain, Chen, is poorly supported (Xiao et al. 1994). According to preliminary data of Arikawa et al. (1999), HTN virus strains from China, Korea, and Japan form several lineages, with Korean and Japanese strains clustering separately from the Chinese ones.

Knowledge about the genetics and evolution of the HTN virus' rodent host, *A. agrarius*, is quite limited at the moment. It was shown that the striped field mice from the Far East and from Europe belong to two distinct subspecies, *A. agrarius mantchuricus* and *A. agrarius agrarius*, respectively (Chernukha et al. 1986). It is noteworthy that HTN virus is found only in the Far Eastern subspecies of *A. agrarius*. As for the whole *Apodemus* genus, its evolutionary age was estimated as approximately 6–12 million years (Chelomina et al. 1998; Chelomina 1998). According to the mtDNA cytochrome b sequences of nine *Apodemus* species, the genus might be divided into three distinct groups: south Asian (that includes *A. argenteus* and *A. semotus*), east Asian (*A. agrarius, A. speciosus*), and Eurocaucasian (*A. flavicollis, A. sylvaticus, A. uralensis, A. ponticus, A. flavipectis*) (Chelomina et al. 1998). The authors, however, admit, that the phylogeny inferred from their sequence data is somewhat different from phylogenies based on other genetic data as well as from the current zoological taxonomy of wood and field *Apodemus* mice.

2.1.2 Seoul Virus

Seoul (SEO) virus is another agent for the HFRS, though it causes a milder form of the disease than that associated with HTN virus. SEO virus' natural hosts, black and Norway rats *Rattus rattus* and *R. norvegicus*, are much more mobile that hosts of other hantaviruses. Traveling by ships, rats are worldwide in distribution, as is the virus. So far, sequence data were obtained for SEO virus strains originating from Asia (Korea, Japan, China), Africa (Egypt), Americas (USA and Brazil) (Xiao et al. 1994), and Europe (Pilaski et al. 1999). Although locally SEO virus strains show geographic clustering of genetic variants, with the M-segment sequence diversity up to 5% between Japanese and Chinese strains (Kariwa et al. 1994), the global pattern remains unclear. It is likely to reflect human trading and migratory routes across seas and oceans. Thus, SEO virus represents an example of global anthropogenic expansion of an infectious agent, and this is the only "cosmopolitan" hantavirus known so far.

2.1.3 Dobrava Virus

Dobrava (DOB) virus was originally discovered in Slovenia. The prototype strain was isolated from the yellow-necked mouse (*A. flavicollis*) trapped near Dobrava village, which gave the name to the hantavirus type (Avsic-Zupanc et al. 1992). DOB virus is connected to a severe form of HFRS in the Balkans (Slovenia, Greece, Albania, Bosnia); although the scale of the infection is much smaller than that of HTN virus, between 9 and 12% of all hospitalized cases are fatal (Papa et al. 1998; Avsic-Zupanc et al. 1999). On the phylogenetic trees, all *A. flavicollis*-derived sequences from this part of Europe are clustered together and are monophyletic with sequences recovered from HFRS patients with confirmed DOB virus infection.

Surprisingly, in Estonia, a hantavirus closely related to DOB virus was found in the striped field mouse (Plyusnin et al. 1997b; Nemirov et al. 1999), the *Apodemus* species known to harbor HTN virus in the Far East. However, as already mentioned, the striped field mice from Europe belong to the distinct subspecies, *A. agrarius agrarius* (Chernukha et al. 1986). The Saaremaa isolate showed 27–29% genetic diversity from the HTN virus and only 12–19% from Slovenian DOB virus. Later, DOB-like virus sequences were recovered from *A. agrarius* trapped in Russia (Plyusnin et al. 1999a) and Slovakia (Sibold et al. 1999b). *A. agrarius*-derived sequences formed two distinct lineages comprising variants from Estonia, on one hand, and from Russia and Slovakia, on the other. Sharing a common ancient ancestor with the *A. flavicollis*-originating lineage, two *A. agrarius*-derived lineages are not necessarily monophyletic: with the current set of sequence data, placing them together does not receive a sufficient bootstrap support, and in some phylogenies the Estonian lineage is even clustered (albeit with low probability) together with the one from the Balkans.

These observations raised a question whether the hantavirus variants harbored by the yellow-necked and the field striped mouse represent geographical variants of DOB virus or belong to distinct hantavirus species (or subspecies). Current data suggest that the latter might be the case, i.e., "classic" DOB virus with Slovenian prototype Dobrava strain (Avsic-Zupanc et al. 1995) could be distinguished from the Saaremaa (SAA) virus, with the Estonian isolate (Nemirov et al. 1999) as a prototype. Indeed, besides having different primary rodent reservoirs, the two viruses show fourfold titer differences in neutralization tests (Nemirov et al. 1999; A. Lundkvist, unpublished data), thus, meeting at least two of the four criteria currently selected to define distinct hantavirus species (types) (Elliott et al. 1999). As for the third criterion (7% diversity for the deduced amino acid sequences), they come close to the borderline with 6.1% diversity for the complete glycoprotein precursor sequence, the nucleocapsid protein sequences show diversity of 3% only. The fourth requirement (rather more theoretical than practically applicable at the moment), the inability of forming natural reassortants with other hantavirus species, remains to be evaluated for this case.

The ranges of the two *Apodemus* species overlap over most of Europe (Mitchell-Jones et al. 1999). However, until very recently neither DOB virus was found associated with *A. agrarius* in the Balkans, nor were sequences belonging to this hantavirus recovered from *A. flavicollis* outside the region. The most recent findings provide additional support for the demarcation between DOB and SAA viruses that stems from the hantavirus-host co-speciation and co-evolution: partial virus sequences recovered from Slovenian *A. agrarius* (Avsic-Zupanc et al. 2000) and Slovakian *A. flavicollis* (Sibold et al., unpublished data), instead of showing geographical clustering, clearly fit the host-dependent pattern.

In the regions where SAA virus is dominating, the clinical outcome of infections seems to be milder than for cases associated with classic DOB virus in the Balkans. Indeed, SAA virus caused an outbreak in central Russia in 1991–1992 when 130 HFRS cases, but no casualties, were recorded (Lundkvist et al. 1997). Similarly, in Estonia, where antibodies to SAA virus are present in 1% of healthy individuals and the first clinical cases have recently been confirmed (Lundkvist et al. 1998a; Golovljova et al. 2000), no associated mortality has so far been registered. Thus, with all reminiscences about lack of clinical alert etc., DOB and SAA viruses seem to possess different pathogenicity for humans, and their detailed comparison might shed some light on markers of pathogenicity.

As mentioned above, the knowledge about both phylogenetic relationships between members of the *Apodemus* genus and their genetic diversity is very limited. According to the mtDNA cytochrome b sequence, *A. flavicollis* belongs to the Eurocaucasian group, distinct from both the east Asian group (that includes *A. agrarius*) and south Asian group (Chelomina et al. 1998). Considering this, one can observe, that the phylogenetic relationships between HTN, DOB, and SAA viruses contradict those between their natural hosts. For instance, HTN and SAA viruses are not monophyletic as might be expected for hantaviruses carried by two subspecies of *A. agrarius* (see Fig. 1). Such a discrepancy in a number of cases was interpreted as evidence for host-switch event(s) in an evolutionary history of a given group of hantaviruses (described below, e.g., for Topografov, Khabarovsk, Puumala and Sin Nombre, New York, Monongahela). Obviously, there is a demand for comparative phylogenetic studies of HTN, DOB, and SAA viruses and their natural hosts, similar to those performed for Sin Nombre (SN) and SN-like viruses and *Peromyscus* species in North America (Morzunov et al. 1998; Monroe et al. 1999).

2.2 Arvicolinae-Borne Hantaviruses

2.2.1 Puumala Virus

Puumala (PUU) virus, discovered in the late 1970s (Brummer-Korvenkontio et al. 1980), is the most common cause of HFRS in Europe. The disease, also referred to as nephropathia epidemica (NE), is usually milder than those associated with HTN virus, or even SEO virus, and the fatality rate is 0.1% (for review, see

BRUMMER-KORVENKONTIO et al. 1999). The virus is harbored by the bank vole (*Clethrionomys glareolus*), which is found in most of Europe excluding the Mediterranean coast and northernmost areas. In Fennoscandia (this collective term is used for Finland and Scandinavia) *C. glareolus* exhibits population cycles of 3–4 years, and so do human HFRS epidemics. In temperate Europe the bank vole populations are much more stable, but in occasional "mast years" (good seed crops of oak and beech) the rodents become abundant giving rise to HFRS outbreaks or epidemics. In Asia *C. glareolus* is found in Russia (southern Siberia from the Ural Mountains to Lake Baikal), northern Kazakhstan, and the Altai and Sayan Mountains (MUSSER and CARLETON 1993). However, no human PUU virus infections from Asia have been reported so far, except for the Asian slope of the Ural Mountains in Bashkortostan, Russia (S.F. Khaiboullina, unpublished data). Accordingly, the sequence information now exists only for European PUU virus strains of either bank vole or human (HFRS-patients) origin.

Genetic analysis of strains from Finland, Sweden, Norway, Russia, Belgium, Germany, and Austria has shown PUU virus as the most variable of the hantavirus species: genetic diversity on the nucleic acid level reaches 20% and 17%, for the coding regions of the M and S segments, respectively, and even higher values (37% and 30%) for their 3′-noncoding regions (LUNDKVIST et al. 1998b).

The pattern of geographical distribution of *C. glareolus* in Europe is quite complex. It seems to reflect a sequence of glaciations and reglaciations followed by mass migrations of mammalian species. The history of postglacial recolonization of Fennoscandia is especially complicated. The most recent migration of bank voles is thought to have occurred after the retreat of the Weichselian ice sheet (8,000–13,000 years ago) when the territory of Fennoscandia was recolonized from presumably two directions by survivors from glacial refugia located somewhere in the central and southern regions of the European plain. One of the migratory routes to the territory of Sweden went from the south via a land bridge connecting the present Denmark with southern Sweden, while another went from the northeast via northern Finland and Russia; the two streams of recolonizers met somewhere in central Sweden along the line that crosses the Scandinavian peninsula in an amazingly straight manner (Fig. 2) (TEGELSTRÖM et al. 1988). Northern and a southern populations of bank voles can be distinguished by mtDNA sequence, which, for the northern variant, seems to be very similar to that of the closely related species, *C. rutilus* (TEGELSTRÖM et al. 1987). Nowadays, the contact zone between the northern and southern populations of the bank vole in central Sweden is still only 50km wide (JAAROLA et al. 1999). Notably, PUU virus strains, which circulate in the bank voles on the two sides of the contact zone in Sweden, belong to two distinct genetic lineages (HÖRLING et al. 1996a; LUNDKVIST et al. 1998b). In addition, the contact zone between the same two mitochondrial DNA types of the bank vole was found recently in Finland and, amazingly, even far away from Fennoscandia – in Russia, within the well-known PUU virus endemic area near the Ural Mountains, Bashkortostan (MORZUNOV et al. 1999). The contact zone in Russia (as we suspect, its currently known part) is located in western Bashkortostan, Udmurtia, and Perm region. It separates the western (mtDNA type similar to that of

Fig. 2. Co-speciation of the two distinct PUU virus lineages with the two genetically distinct populations of the bank vole in Fennoscandia. The shaded area crossing Scandinavia and Finland represents the contact zone between the two bank vole populations, which are distinct in respect of their mtDNA. White and black "mouse" figures designate localities containing "northern" or "southern" mtDNA haplotypes, respectively (S.P. Morzunov and A. Plyusnin, unpublished data). White and dark "virion" figures indicate localities where "northern" or "southern" Scandinavian and Finnish PUU virus lineages were recovered from *C. glareolus* (Hörling et al. 1996a; Lundkvist et al. 1998b)

the "southern" bank voles from Fennoscandia) and the eastern ("*C. rutilus*-like" mtDNA type similar to that of the "northern" bank voles from Fennoscandia) populations of *C. glareolus*. These new findings suggest that postglacial migrations of the bank vole in Eurasia took place at a much greater scale than was previously thought, and that more data should be collected from the entire contemporary geographic range of the bank vole to facilitate further investigation of the historic relationship of this rodent species with the Puumala virus.

Most recent phylogenetic analyses show five distinct lineages of European PUU virus: Finnish, Russian, Belgian, and two Scandinavian (strains from northern Sweden, and Central Sweden and Norway, respectively). The first Scandinavian PUU virus lineage is represented by strains from Vindeln, Hällnäs, Mellansel, and Hundberget, all located north of the contact zone. The second lineage consists of strains from Sollefteå and Sundsvall (Sweden) and from Eidsvoll and Vinstra (Norway), all located south of the population border, thus confirming co-speciation pattern and separate phylogenetic history of the "northern" and "southern" Scandinavian PUU virus strains (Fig. 2). A similar pattern of the phylogenetic branching is observed among Russian PUU virus strains. The first group of strains (Giebel et al. 1989; Stohwasser et al. 1990; Xiao et al. 1993b) is genetically related to the original Russian rodent-derived strain CG18–20 from Bashkortostan (pre-Ural area), and the second group comprises the strains found in Udmurtia and Tatarstan (Plyusnin et al. 1994a, 1995a). These two groups display up to 15% M-genome segment difference on the nucleic acid level, while corresponding genetic divergence within each of these groups usually does not exceed 4–6%. Interestingly, our recent data seem to confirm that these two phylogenetic clades of the PUU virus in Russia are located, respectively, east and west of the contact zone mentioned above (KHAIBOULLINA et al. 1997; MORZUNOV et al. 1999). In addition, genetic reassortment between the western (Tatarstan/Udmurtia) and eastern (Bashkortostan) PUU virus strains was detected in eastern Bashkortostan, with the "western type" S-genome segment widely invading populations of the eastern PUU virus in central Bashkortostan (KHAIBOULLINA et al. 1997).

The Finnish PUU virus lineage comprises the prototype strain Sotkamo (VAPALAHTI et al. 1992), strain Puumala (from the very same bank vole in which the virus was originally discovered), and other bank vole and human originating strains from that country (PLYUSNIN et al. 1995a, 1997a, 1999b). The Belgian lineage of PUU virus is represented by the strain Cg13891 originating from the northern part of the country (BOWEN et al. 1995; ESCUTENAIRE et al. 2000). Partial sequences reported for strains from Germany (PILASKI et al. 1994) and Austria (BOWEN et al. 1997) represent two other lineages.

Recently, sequences of PUU-like virus were recovered from tissue samples of *C. rufocanus* trapped in Hokkaido (Japan) (KARIWA et al. 1999). On the phylogenetic tree, two *C. rufocanus*-originating strains, designated Tobetsu and Kamiiso, cluster in the closest proximity to the European *C. glareolus*-originating strains and share with them a common ancient ancestor. As this PUU-like virus has not been isolated in cell culture yet, its serological and other properties remains undetermined. Interestingly, unlike European PUU virus, it appears to be nonpathogenic for humans: no positives were found in a recent immunologic survey of over a thousand human sera from Hokkaido (KARIWA et al. 2000).

In general, the phylogeny of PUU virus looks very "star-like": each of the distinct lineages is well-supported, but interrelationships between them are obscure. The probable explanation is that the lineages were radiating from a common ancestor rather independently. It is of special interest that the PUU-like Tobetsu virus is the nearest to the ancestral node separating the PUU virus clade from other hantaviruses. Probably, a split of PUU and Tobetsu types was caused by a divergence of the "ancestral PUU virus" into *C. glareolus*- and *C. rufocanus*-specialized variants. Recently, partial S- and M-genome segment sequences of yet another PUU-like virus strain, Muju virus, were amplified from the vole *Eothenomys regulus*, a Korean endemic that was considered a subspecies of *C. rufocanus* until 1978 (SONG et al. 1999). Findings of the divergent PUU-like virus lineages in two vole species related to the natural host of PUU virus (*Clethrionomys glareolus*) are consistent with co-speciation. However, further study will be needed to clarify this evolutionary pattern for the PUU-like viruses.

2.2.2 Microtine-Borne Hantaviruses from the Old World

On the phylogenetic trees, microtine-borne hantaviruses from the Old World, Tula (TUL) and Khabarovsk (KBR) viruses, are clustered together with other hantaviruses carried by arvicoline rodents. However, while TUL virus is monophyletic with Prospect Hill (PH), Bloodland lake (BLL) and Isla Vista (ILV) viruses (all carried by *Microtus* species from the New World), the KBR virus is paraphyletic to this quartet, and instead shares an ancestor with Topografov (TOP) and PUU viruses (Fig. 1). This is indicative of possible host-switch events in its evolution (see below).

TUL virus was first discovered in tissue samples of the European common vole (*Microtus arvalis*) trapped in central Russia (Tula is a town located 180km south of Moscow) (Plyusnin et al. 1994b).[4] Later, this hantavirus was found in *M. arvalis* in Moravia (Czech Republic) (Plyusnin et al. 1995b), Slovakia (Sibold et al. 1995, 1999a), Austria (Bowen et al. 1997), Poland (Song et al. 1998), and Switzerland (F. Saucy and A. Plyusnin, unpublished observations). As the European common vole is widely spread in Central and Eastern Europe (Mitchell-Jones et al. 1999), one can expect to find TUL virus in other European countries.

TUL virus seems to be nonpathogenic for humans, resembling in this respect its cousins from the New World, PH, ILV, and other PH-like viruses; screening of hundreds of human sera from Russia and Moravia revealed only one TUL virus-positive, and that person, according to the medical records, had no HFRS- or NE-like disease (Vapalahti et al. 1996; A. Plyusnin et al., unpublished data).

On the phylogenetic tree based on the complete S-segment coding sequences, TUL virus strains are monophyletic and form two distinct clades. The first clade includes two lineages formed by strains from Russia and east Slovakia, while the second clade includes strains from west Slovakia, Moravia, and Switzerland. According to the partial sequences known for TUL virus strains from Austria (Bowen et al. 1997), they can be placed within the second clade. Interesting observations were made using detailed genetic analysis of TUL virus strains from east Slovakia (Sibold et al. 1999a). The contradictory phylogenies inferred from different parts of the S segment as well as a mosaic pattern of amino acid substitution in the deduced sequences of the nucleocapsid protein suggested that this lineage represents, in fact, a product of a homologous recombination between the two lineages of TUL virus. Earlier data on phenotypic classification of European common vole (Kratochvil 1959) suggested that during the last reglaciation, the territory of former Czechoslovakia was re-colonized by *M. arvalis* from three directions: from the west, by *M. arvalis arvalis*, from the east, by *M. arvalis duplicatus*, and from the south, by *M. arvalis levis*. Thus, representatives of distinct TUL lineages might be brought together in this part of Europe with different migration streams and then recombine yielding in a new lineage. Similar to the case of SAA virus, studies on TUL virus' rodent host genetics would be very useful in further evaluating this hypothesis.

---

[4] In fact, the TUL virus S-segment sequence was recovered also from one tissue sample of *Microtus rossiaemeridionalis*, the species that is sibling to *M. arvalis*. These two species could only be differentiated by chromosomal analysis. Whether this *Microtus* species is an alternative natural host for TUL virus remains to be determined.

2.2.3 Topografov Virus

TOP virus was first discovered in tissue samples of Siberian lemmings (*Lemmus sibiricus*), trapped in the basin of the Topografov river in Taimyr Peninsula (Plyusnin et al. 1996b). This area is populated by the western type of *L. sibiricus*, which is found in the Siberian tundra along the arctic coast up to Taimyr in the East. In fact, the western type of *L. sibiricus* is more closely related to the Norwegian lemming of Fennoscandia (*Lemmus lemmus*) than to the central type of *L. sibiricus*, inhabiting the huge area between the Lena and Kolyma rivers in Siberia, or the eastern type (known also as *L. trimucronatus*), distributed east of Kolyma River and in Alaska (Fedorov et al. 1999). *L. lemmus* is famous for its mass migrations caused by drastic fluctuations of population density (Henttonen and Kaikusalo 1993). Such a dispersal capacity of lemmings is probably responsible for the observed lack of clear geographic clustering of genetic variants, when the wild type TOP virus strains from two localities, which are about 600km apart, were analyzed (Vapalahti et al. 1999). In the first locality, two genetic lineages of TOP virus, diverged by 7.5% in the S-segment sequences, were co-circulating, and, surprisingly, one of the lineages showed a closer relatedness (1% nt divergence) to the TOP virus strains from the other locality.

Recently, TOP virus was isolated in Vero E6 cells and in laboratory-bred Norwegian lemmings (Vapalahti et al. 1999). In the related paper, an attempt was made to compare phylogenies inferred from the hantaviral sequences and those of mitochondrial DNA from the host rodent species. Although the conclusion was that the co-evolution of hantaviruses and their hosts is a general rule, some exceptions were observed. The most prominent of them is the grouping of KBR virus, the hantavirus found in the reed voles (*Microtus fortis*) in Russia's far east (Dzagurova et al. 1995; Hörling et al. 1996b), separately from other *Microtus*-borne viruses, but instead with TOP virus, carried by *Lemmus*. Moreover, the TOP/KBR pair shares a common ancestor with PUU virus, and its branching from the PUU virus lineage is distal to the main *Microtus*-derived virus cluster comprising PH, TUL, BLL, and ILV viruses (Fig. 1). It is, again, contradictory to the phylogeny of the rodent hosts since *Clethrionomys* and *Microtus* species are more closely related to each other than to *Lemmus*, whose position on the tree is more ancestral. Thus, the location of an ancestor for TOP and KBR viruses is not in line with the phylogeny with either of the host rodents and might be explained by a sequence of two host-switch events: an ancestor of *Clethrionomys* "donated" to *Lemmus* a virus which became the common ancestor of both TOP and KBR viruses; then, after some time of co-evolution/co-speciation, a "jump" to *Microtus* occurred. Although time points for both events remain unknown, the second one should be rather recent, as the two virus species, TOP and KBR, are still radiating. It reflects, for example, in the fact that the 3′-noncoding regions of KBR and TOP virus S and M segments are alignable (a unique case for hantaviruses), except for 113 and 58 nt deletions, respectively, in KBR.

TOP virus M segment showed 77% nucleotide (nt) and 88% amino acid (aa) identity with KBR virus and 76% nt and 82% aa identity with PUU virus.

However, the identity between TOP and KBR virus S segments was disproportionately higher: 88% at the nt and 96% at the aa level (Vapalahti et al. 1999). One plausible explanation for this is a reassortant nature of the ancestral virus, which could be responsible for its higher "host-switch potential". Alternatively, the difference could mean that, after the host-switch, an evolutionary pressure in a new host generated more amino acid changes in the glycoproteins than in the nucleocapsid protein.

3 New World Hantaviruses and Their Rodent Hosts

As mentioned above, hantaviruses are associated with three groups of murid rodents (subfamilies Murinae, Arvicolinae, and Sigmodontinae). There are two phylogenetically distinct groups of New World sigmodontines: neotomine-peromyscine (18 genera and 124 species), which are found predominantly in North America, and South American sigmodontines (61 genera and 299 known species) (Musser and Carleton 1993). One or more genera of sigmodontines are believed to have arrived in South America between 9 and 3 million years ago (reviewed in Engel et al. 1998). A rapid adaptive radiation that occurred following the original invasion seems to account for the high genetic diversity of the contemporary South American sigmodontines.

However, due to the absence of human epidemics of recognizable hantavirus disease, the scientific world was for a long time unaware of the existence of hantaviruses in indigenous New World rodents. Even by the late 1980s, PH virus, believed to be nonpathogenic for humans and associated with the arvicoline *Microtus pennsylvanicus* (the meadow vole), was the only recognized indigenous New World hantavirus (Parrington and Kang 1990; Parrington et al. 1991). In 1993, SN virus (Nichol et al. 1993), hosted by the sigmodontine *Peromyscus maniculatus* (the deer mouse) (Childs et al. 1994), was discovered to be the cause of the often-fatal HPS (Duchin et al. 1994). Since that time, advances in PCR technology and molecular diagnostics facilitated genetic characterization of several novel New World hantaviruses associated with different species of sigmodontine and arvicoline rodents (Hjelle et al. 1994a,b, 1995a,b,c; Khan et al. 1995; Morzunov et al. 1995, 1998; Ravkov et al. 1995; Rollin et al. 1995; Rowe et al. 1995; Song et al. 1995, 1996; reviewed in Nichol et al. 1996; Plyusnin et al. 1996a; Schmaljohn and Hjelle 1997; Monroe et al. 1999; Nichol 1999).

3.1 Sigmodontinae-Borne Hantaviruses

3.1.1 Sin Nombre-Like Viruses Co-Speciate with Their Rodent Hosts

Classic western SN virus (Nichol et al. 1993; Elliott et al. 1994; Hjelle et al. 1994a; Spiropoulou et al. 1994; Chizhikov et al. 1995; Schmaljohn et al. 1995) was discovered to be the causative agent of an outbreak of HPS (Duchin et al. 1994) in the southwestern United States in 1993. The deer mouse (*P. maniculatus*) was quickly identified as the primary natural reservoir for SN (Childs et al. 1994). An extensive study of SN virus throughout North America revealed a complex picture of high genetic diversity as well as geographic clustering of genetic variants (Hjelle et al. 1994c; Spiropoulou et al. 1994; Henderson et al. 1995; Rowe et al. 1995; Monroe et al. 1999). Additionally, several other genetically divergent hantavirus genotypes (referred to as SN-like viruses) have been detected and genetically characterized. Monongahela (MON) virus (Song et al. 1996) has been described originating from a distinct subspecies of *P. maniculatus*, *P. m. nubiterrae* (cloudland deer mouse), from the Appalachian Mountains. New York (NY) virus (Hjelle et al. 1995b,c) and Blue River (BR) virus (Morzunov et al. 1998; Monroe et al. 1999) were recovered from tissue samples of *Peromyscus leucopus* (white-footed mouse), which were collected in the coastal areas of Rhode Island/New York and in several locations within Great Plains, respectively. The existence of several distinct SN-like virus genotypes within the geographic ranges of two closely related rodent species, *P. maniculatus* and *P. leucopus*, and the geographic partitioning of SN-like viruses of *P. maniculatus* raised questions concerning the long-term evolution and maintenance of these viruses in the populations of their rodent hosts, and possible co-speciation of SN-like viruses with particular *Peromyscus* taxa.

Mice of the genus *Peromyscus* (Muridae; Sigmodontinae), occur in North America in nearly all terrestrial habitats from the Atlantic to Pacific seaboards and from the Canadian Northwest Territories to Panama (Baker 1968; Hall 1981). *P. maniculatus* not only has the most extensive range within the genus (covering most of the United States and Canada), but is also the most morphologically and ecologically diverse with more than 65 subspecies currently recognized (Osgood 1909; Baker 1968; Hooper 1968; Hall 1981; Hogan 1993). Seventeen subspecies of *P. leucopus* occupy the eastern half of the United States (Osgood 1909; Baker 1968; Hooper 1968; Hall 1981), and, in most of their habitats, are sympatric with *P. maniculatus*. Despite considerable efforts, intraspecific phylogenetic histories of both species remain unclear (reviewed in Carleton 1989). Traditionally, *P. maniculatus* has been divided into two major morphologic and ecologic types, each embracing many subspecies (Osgood 1909; Blair 1950; Hooper 1968). A larger long-tailed type (forest form) inhabits forested regions of the Appalachians from northern Georgia and northward to Labrador, across the continent in the Canadian forests, and southward through the Rocky Mountains and mountains of the Pacific coast. A smaller, short-tailed, small-eared, small-footed type (grassland form) occupies the interior prairies and grasslands of the continent and extends into deserts of the western and southwestern United States. In the case of *P. leucopus*, the existence of two distinct chromosomal races, Northeastern and Southwestern, has been well documented during the last two decades (Baker et al. 1983; Stangl 1986; Nelson et al. 1987; Van Den Bussche et al. 1993). Chromosomal, allozymic, and mtDNA restriction pattern data suggest different phylogenetic histories for the two chromosomal races and an allopatric origin of the geographically stable narrow hybrid zone found in Oklahoma (STANGL 1986; NELSON et al. 1987; VAN DEN BUSSCHE et al. 1993).

As more data on the SN-like viruses emerged, it became clear that the genetic diversity and geographic distribution of these viruses are related to the genetic diversity, geographic distribution, and phylogenetic history of their rodent hosts. Molecular approaches involving comparisons of phylogenetic lineages inferred from genetic sequences of SN-like virus genotypes and their corresponding rodent hosts appeared to be an extremely useful tool to address these issues (MORZUNOV et al. 1998). Phylogenetic analysis of the numerous mitochondrial DNA (mtDNA) sequences of *P. maniculatus* collected throughout its geographic range revealed four well-supported major clades (MORZUNOV et al. 1996). Geographically these clades correspond to central, northeastern, northwestern, and southwestern portions of the deer mouse distribution, with the first two clades containing the majority of the samples analyzed. The central clade primarily contains sequences from mice representing subspecies traditionally recognized as the "grassland form" of *P. maniculatus*. The northeastern clade comprises mtDNAs for the *P. maniculatus* "forest form" subspecies *abietorum*, *gracilis*, and *nubiterrae* from eastern United States and Canada (Fig. 3). Two remaining minor clades comprise mtDNA sequences for the "forest form" deer mice from the Pacific Northwest, and from the Sierra Nevada Mountains and Baja California, respectively.

When rodent host phylogeny was compared with phylogeny of SN-like viruses throughout North America, it became clear that two of these virus genotypes, classic western SN and MON viruses, are restricted to the geographic ranges of the phylogenetically (and morphologically) distinct forms of *P. maniculatus*, the grassland form and the eastern forest form, respectively (Fig. 3). This finding is consistent with the hypothesis that hantaviruses "co-speciate" (i.e., diverge in close concert) with their specific rodent hosts. The co-existence of two contrasting lineages of the classic western SN virus in *P. maniculatus* in Nevada and California (HENDERSON et al. 1996) may also indicate even finer co-speciation on the population level, through geographic separation and following re-unification of the corresponding deer mouse populations. However, since these two contrasting genotypes are both apparently hosted by the grassland form deer mice, mtDNA analysis does not provide enough phylogenetic resolution to distinguish host populations involved, and draw conclusions about historic migratory routes of these mice (S.P. Morzunov, unpublished data).

P. leucopus mtDNA sequences were found to form four separate genetic clades, referred to here as Eastern, Central, Northwestern, and Southwestern groups (MORZUNOV et al. 1998). The first clade contains mtDNA sequences from the eastern United States including the coastal area, the Appalachian Mountain region, and areas between the Appalachians and Great Lakes. The second clade comprises mtDNA sequences from the northern region of the Great Plains, and the third clade contains mtDNA sequences from Kansas and Arkansas. The fourth clade comprises sequences from Texas and Oklahoma and apparently

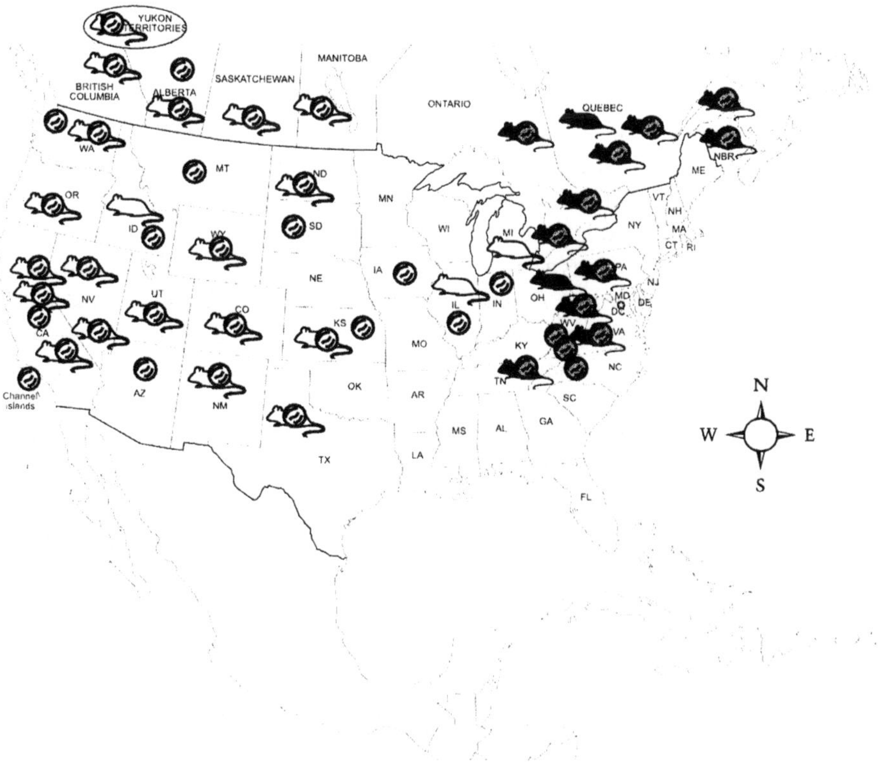

Fig. 3. Co-speciation of the two distinct SN-like viruses with the two genetically distinct forms of the deer mouse in North America. White and black "mouse" figures designate localities containing "central" or "northeastern" deer mouse mtDNA haplotypes, respectively (MORZUNOV et al. 1996; S.P. Morzunov, unpublished data). These two phylogenetically distinct mtDNA haplotypes appeared to be restricted to the geographic ranges of the two known morphologically distinct forms of *P. maniculatus*, the grassland form and the eastern forest form, respectively (reviewed in CARLETON 1989). White and dark "virion" figures indicate localities where classic SN virus or MON virus were recovered from *P. maniculatus* and/or humans (according to MONROE et al. 1999)

corresponds to the known Southwestern chromosomal race of *P. leucopus* (BAKER et al. 1983).

Comparison of the data on *P. leucopus* mtDNA phylogeny with that on phylogeny of *P. leucopus*-borne SN-like virus genotypes strongly indicated that two phylogenetically closely related, though divergent, virus genotypes from Indiana and Oklahoma (recently termed Blue River virus, BR) (MORZUNOV et al. 1998; MONROE et al. 1999; NICHOL 1999) are associated with two phylogenetically distinct mtDNA haplotypes (and distinct contemporary chromosomal races) of the white-footed mouse. Once again, a clear evolutionary pattern of co-speciation of BR virus lineages with phylogenetically distinct populations of the host species can be seen. The BR virus lineage from Oklahoma is clearly associated with Southwestern chromosomal race (and the Southwestern mtDNA haplotype) of

P. leucopus. The virus genotype from Indiana is associated with the Northeastern chromosomal race (Eastern or Northwestern mtDNA haplotype)[5].

Surprisingly, detailed phylogenetic analysis of the SN-like virus sequences (Morzunov et al. 1998) clearly show that *P. leucopus*-borne NY virus (Hjelle et al. 1995b,c), which was found in white-footed mice of the Eastern mtDNA haplotype in coastal areas of Rhode Island and New York, was most closely related not to BR, but to the *P. maniculatus*-borne viruses, SN and MON, and monophyletic with MON virus. The most likely explanation for the unusual position of NY virus on the hantavirus phylogenetic tree is a single historically recent host-switching event. Both *P. maniculatus nubiterrae* and *P. leucopus* are known to inhabit the Appalachian Mountain region, often sharing common habitats (Baker 1968; Hall 1981; Wolff 1986, 1989). This situation is thought to have existed since the last (Wisconsin) glacial period when this region served as an ice age refuge for both species (Waters 1963). Most likely "an ancestral MON-like virus" was originally associated with *P. maniculatus*, based on virus position on the phylogenetic tree and its current association solely with *P. maniculatus* in Quebec and Ontario, outside of the geographic range of *P. leucopus* (Monroe et al. 1999). At present, this virus can be found in the southern Appalachians predominantly in *P. maniculatus nubiterrae*, but also in *P. leucopus* (Monroe et al. 1999), suggesting that "ancestral MON virus" may have become well-adapted to both closely related rodent species during the ice age. After the glacial ice sheet retreated from New England more than 10,000 years ago, ancient populations of both species expanded northward along the Appalachian Mountain chain. It seems likely that co-speciation of the ancient MON-like virus genotype with coastal populations of *P. leucopus*, which migrated out of range of *P. maniculatus*, resulted in the constitution of the SN-like virus genotype known as NY virus.

3.1.2 Other North American Sigmodontine-Borne Hantaviruses: Further Concordance of Virus and Rodent Phylogenies

Shortly after the original discovery of SN virus, hantaviruses were serologically detected in several other indigenous North American rodent species. Phylogenetic analysis of the RT-PCR amplified virus sequences revealed two additional groups of the North American hantaviruses, which display at least 28% M-genome segment nucleotide differences (see the table in Morzunov et al. 1998) between them and from the group of SN-like viruses. The first group is associated with rodent genus *Reithrodontomys*. This genus represents another branch of neotomine-peromyscine rodents, and it is placed as a sister taxon to genus *Peromyscus* in recent phylogenetic

[5] Since BR virus genotype from Indiana was discovered at the western edge of the geographic distribution of white-footed mice of the Eastern mtDNA haplotype, it is not clear at this point whether white-footed mice possessing Eastern or, alternatively, Northwestern mtDNA, or both, harbor this virus genotype. Based on the absence of this virus lineage in samples from the rest of the geographic range of *P. leucopus* of the Eastern mtDNA haplotype (fairly well-sampled Appalachians and coastal areas) (Hjelle et al. 1995c; Song et al. 1996; Monroe et al. 1999), it seems likely that this virus lineage is predominantly associated with *P. leucopus* of the Northwestern mtDNA haplotype (which inhabits areas west from Indiana) and invaded "border" populations of *P. leucopus* of the Eastern mtDNA haplotype.

reconstructions (ENGEL et al. 1998). *Reithrodontomys megalotis* (western harvest mouse), the most common *Reithrodontomys* species in the western United States and Mexico, was found to harbor El Moro Canyon (ELMC) virus (HJELLE et al. 1994b). A related virus genotype (22% S-genome segment nucleotide difference), termed Rio Segundo (RIOS), was discovered in Costa Rica, and is most likely associated with *Reithrodontomys mexicanus* (Mexican harvest mouse) (HJELLE et al. 1995a).

The second group of North American sigmodontine-borne viruses is associated with *Sigmodon hispidus* (hispid cotton rat) and *Oryzomys palustris* (marsh rice rat). These rodent genera (*Oryzomys* and especially *Sigmodon*), are considered to be basal branches of South American sigmodontines (ENGEL et al. 1998). Different species of cotton rats and rice rats (and hantaviruses associated with them) can be found in the southern United States, Mexico, Central America, and South America (MUSSER and CARLETON 1993). Two closely related virus genotypes, Black Creek Canal (BCC) virus (RAVKOV et al. 1995; ROLLIN et al. 1995) and Muleshoe (MUL) virus (RAWLINGS et al. 1996), are associated with hispid cotton rat in the eastern and the western half of its geographic range in the southern Unites States, respectively. There is a strong indication that the western and the eastern populations of *Sigmodon hispidus* represent morphologically and phylogenetically distinct subspecies (RAWLINGS et al. 1996), thus, providing another example of co-speciation of the virus genotypes with corresponding populations of its rodent host species. As one could expect, the next closest relative of these North American *Sigmodon*-borne viruses is Caño Delgadito (CD) virus (FULHORST et al. 1997) isolated from South American *Sigmodon* species, *Sigmodon alstoni*, from Venezuela[6]. Bayou (BAY) virus, a distinct virus genotype associated with the marsh rice rat, represents the last currently known member of this clade (MORZUNOV et al. 1995; KSIAZEK et al. 1997; TORREZ-MARTINEZ et al. 1998). Interestingly, recent phylogenetic analysis of the entire virus S-genome segment places BAY, but not MUL, as the closest relative of BCC (NICHOL 1999). This could be suggestive of yet another host-switching event. However, further studies including phylogenetic analysis of the M-genome segments of these viruses are needed to clarify this situation.

3.1.3 South American Hantaviruses: Paradise for Virologists – Nightmare for Systematists

The South American continent presents a unique and challenging opportunity for the study of the relationships of hantaviruses with their natural rodent hosts. As the result of a historically recent invasion of sigmodontine rodents and subsequently extremely rapid adaptive radiation (HERSHKOVITZ 1966; SIMPSON 1969; WEBB 1976;

[6] There is currently a discrepancy between CD virus phylogenies reconstructed using M and S RNA segment sequences: while phylogenetic analysis of the M segment clearly places this virus within the clade of *Sigmodon*-borne viruses, analysis of the S segment consistently groups it with *Reithrodontomys*-borne viruses (FULHORST et al. 1997; NICHOL 1999). Interestingly, *Reithrodontomys* is one of only two genera of neotomines-peromyscines that extend into northern South America (MUSSER and CARLETON 1993). Known examples of hantavirus laboratory contaminations and dual infections in cell culture urge us to exercise caution. However, if further proven, this case could be the first example of the natural S-segment genetic reassortment between hantaviruses harbored by different tribes of sigmodontines.

MARSHALL 1979; reviewed in ENGEL et al. 1998), the contemporary genetic diversity of sigmodontines in South America is substantial (MUSSER and CARLETON 1993). The majority of these rodent species have not been well characterized yet, and their association with the indigenous hantaviruses is still poorly understood. However, it is quite clear that ancestral South American hantaviruses did not miss their "golden opportunity" to diversify: in the last 4 years at least ten phylogenetically distinct virus genotypes harbored by South American sigmodontines have been characterized, and several of them isolated in cell culture. Most of these viruses are known already to cause human disease. Representatives of at least three major South American sigmodontine tribes – Phyllotini, Akodontini, and Oryzomyini – appear to harbor hantaviruses. Laguna Negra (LN) virus (JOHNSON et al. 1997; WILLIAMS et al. 1997) is associated with the phyllotine rodent *Calomys laucha*. Pergamino (PGM) and Maciel (MAC) viruses are hosted by akodontine rodents *Bolomys obscurus* and *Akodon azarae*, respectively (LEVIS et al. 1998). Finally, the oryzomyine-borne hantaviruses currently include Rio Mamore (RM) virus associated with *Oligoryzomys microtis* (BHARADWAJ et al. 1997; POWERS et al. 1999), Andes (AND) virus associated with *Oligoryzomys longicaudatus* (LOPEZ et al. 1996, 1997; LEVIS et al. 1998; PADULA et al. 1998, 1999; TORO et al. 1998), Lechiguanas (LEC) virus associated with *Oligoryzomys flavescens* (LEVIS et al. 1998), and several other related to this virus genotypes harbored by other oryzomyine species (LEVIS et al. 1998; PADULA et al. 1999) (see Fig. 1). All currently known South American hantavirus genotypes are monophyletic and share from 75% to 90% S- and M-genome segment nucleotide sequence identity (see Table in LEVIS et al. 1998).

Recent phylogenetic reconstructions, based on the mitochondrial DNA sequences, demonstrated that each of the three rodent tribes mentioned above is likely to be paraphyletic, including derived as well as relatively primitive rodent genera (SMITH and PATTON 1993; ENGEL et al. 1998). This fact makes precise matching of virus and rodent phylogenies difficult until corresponding mtDNA sequences of South American rodent species that carry particular hantaviruses are determined and their phylogenetic placement clarified. In addition, many of these rodent species are sympatric, thus providing opportunities for spillover infections and host-switching events. For instance, tribe Oryzomyini is currently thought to be a paraphyletic group, which includes primitive rodent genera (such as genus *Oryzomys*), as well as a relatively derived, species-rich, and genetically diverse genus *Oligoryzomys* (CARLETON and MUSSER 1989; DICKERMAN and YATES 1995). As expected, all *Oligoryzomys*-borne hantavirus genotypes are monophyletic, while the *Oryzomys*-borne BAY virus is placed with hantaviruses of sigmodons (see Sect. 3.1.2). Within the clade of *Oligoryzomys*-borne viruses, further examples of co-speciation, as well as potential host-switching events, can be found. For instance, we recently reported that geographically separated northern and southern populations of *Oligoryzomys longicaudatus* in Argentina carry two phylogenetically distinct hantavirus genotypes, Oran (ORN) and AND, respectively (LEVIS et al. 1998). Preliminary data, obtained in our laboratory, show that rodents from these two rodent populations also differ in respect of their mtDNA, thus indicating potential co-speciation (S.P. Morzunov, unpublished data). In agreement with the phylogenetic placement of their rodent hosts, *Bolomys*-borne PGM virus and *Akodon*-borne MAC virus are positioned outside of the clade of *Oligoryzomys*-borne viruses (Levis et al. 1998). As an example of a potential host-switching event, *Calomys*-borne LN virus is found in the clade of the *Oligoryzomys*-borne viruses as a sister taxon to RM virus (Nichol 1999).

Thus, the emerging picture of the phylogenetic relationship of South American hantaviruses and their primary rodent carriers appears to be quite complex. In fact, some authors expressed doubts about the specificity of the hantavirus/rodent host associations based on the amplification of several hantavirus sequences from the "wrong" rodent species (Padula et al. 1999; P. Padula, personal communication). However, such a conclusion seems to be highly debatable due to the limited number of virus samples obtained from such "secondary" hosts. There is no doubt that scores of novel virus genotypes associated with other species of South American sigmodontines will be found in the future. Considering the close phylogenetic relationship of numerous South American hantavirus genotypes, the possibility of host switching, and uncertainties in rodent taxonomy, reconstructing in detail the co-evolutionary paths of these viruses and their hosts will be a formidable task.

3.2 Microtine-Borne Hantaviruses: Numerous PH-Like Virus Genotypes in North American *Microtus* Species

Currently, there are several known genetically distinct PH-like virus genotypes, which share from 80% to 90% nucleotide sequence identity, that are associated with different species of North American *Microtus* voles. Besides the original PH (harbored by *M. pennsylvanicus*) (Parrington and Kang 1990; Parrington et al. 1991), only a few of those were given separate names. ILV virus and Bloodland Lake (BLL) virus were found in the California vole, *Microtus californicus* (Song et al. 1995), and the prairie vole, *Microtus ochrogaster* (Schmaljohn and Hjelle 1997), respectively. Several divergent PH-like virus genotypes were found in the California–Nevada area (Rowe et al. 1995) and elsewhere throughout the United States (Monroe et al. 1999). Most of those are likely to represent variants of three major vole-associated viruses mentioned above. A few sequences that were derived from another vole species, *Microtus montanus* (montane vole) (Rowe et al. 1995; Monroe et al. 1999), could represent a novel PH-like virus specifically associated with this rodent species. As PH-like viruses are currently considered nonpathogenic for humans, this group has not been studied in detail. Based on the limited data available, it is difficult at this point to make definitive conclusions about the extent of the genetic variability of the *Microtus*-borne PH-like viruses, the specificity of the virus/host associations, and the frequency of spillover infection among different vole species. Nevertheless, available data support a co-evolutionary scenario.

4 Hantaviruses and Their Hosts as Co-Evolving Systems

4.1 Mode of Evolution and Factors Affecting Hantavirus/Host System

The genus *Hantavirus* is rather unique within family *Bunyaviridae*, in which other four genera are made of arboviruses (ELLIOTT et al. 1999). It is beyond the scope of this review to speculate on whether absence of the arthropod vector in the virus lifestyle is an ancient or a derived feature within *Bunyaviridae*. Nevertheless, it is this particular feature that seems to be crucial in the evolution of hantaviruses and predetermined observed patterns of the tight association of these viruses with their specific rodent hosts. Having the arthropod vector (especially flying vector) in the virus life cycle facilitates infection of the mammalian host species on the territory randomly (or semi-randomly) with the entire pool of available local virus strains, improves efficiency of transmission (inoculation of the virus directly into the blood), and facilitates transmission of the virus to possible alternative hosts. In addition, arboviruses ought to have more flexible molecular machinery, which has to function efficiently in the different genetic environments of both their vectors and mammalian hosts.

Absence of the arthropod vector, combined with moderate efficiency of rodent-to-rodent horizontal transmission of hantaviruses in natural settings, severely limits virus ability to spread fast throughout large geographic areas and move between different host species, and, therefore, effectively restricts virus evolution to the local populations of the rodent host. As speciation of the given host population proceeds, hantavirus responds by further adapting to the changing genetic environment of its host. Through millions of years of specific association of hantaviruses with rodents, such co-evolution resulted in remarkably congruent phylogenetic trees for viruses and their hosts (Fig. 1). Similar patterns of branching, reflecting co-speciation, can be observed from the basal nodes of these trees (hantaviruses harbored by murine, arvicoline, and sigmodontine rodents form three distinct phylogenetic clades, irrespective of their geographic distribution) to the very fine terminal branches (specific association of SN-like viruses with genetically distinct forms of the deer mouse and the white-footed mouse). Apparently, hantaviruses were not always successful in following speciation of their hosts: while some virus lineages were able to persist in speciated hosts, some others became extinct. Examples of such sorting events can be found in North American rodent genus *Peromyscus*: while both *P. maniculatus* and *P. leucopus* harbor SN-like viruses, two of their peripheral sister species, *Peromyscus polionotus* and *Peromyscus gossypinus*, respectively, appear to be virus-free.[7] As seen in Fig. 1, clear pattern of co-speciation could be occasionally disrupted by host-switching events.

[7] As noticed more than a decade ago (GAVRILOVSKAYA and BOIKO 1986), hantaviruses are primarily harbored by abundant, widespread rodent species, which usually dominate local rodent communities. It leads one to suggest that there is a feature in the hantavirus lifestyle that favors their maintenance in the large continuous host populations and makes these viruses prone to extinction when host populations experience genetic "bottlenecks".

As for now, three examples of host-switching by hantaviruses have been investigated: (a) association with *Microtus fortis* of KBR virus, which is monophyletic with TOP virus of lemmings (as noted above, KBR/TOP branch, in turn, displays a strong affinity to the PUU virus phylogenetic clade, thus, suggesting additional "original" host-switching of the ancestral PUU-like virus from voles to *Lemmus* sp.) (Vapalahti et al. 1999); (b) association of SAA virus, which is closely related to DOB virus of *Apodemus flavicollis*, with the western form of *Apodemus agrarius* (Plyusnin et al. 1997b; Nemirov et al. 1999) (eastern form of *A. agrarius* has been known for a long time as a primary host of HTN virus in Asia); (c) association with *P. leucopus* of NY virus, which is phylogenetically closer to SN-like viruses of *P. maniculatus* (Morzunov et al. 1998). These examples prove that hantaviruses sometimes can cross the host-species barrier and establish new virus/host systems. However, host-switching does not appear to be a common place among hantaviruses even under conditions of sympatricity of potential rodent hosts, despite a high frequency of spillover infections. For instance, although *P. maniculatus* and *P. leucopus* are sympatric almost everywhere in the North American Great Plains (including Indiana and Oklahoma), the BR virus lineages have not been detected in *P. maniculatus* captured in this region despite analysis of numerous specimens (Monroe et al. 1999). Another example of the association of the hantavirus with only one of closely related sympatric rodent species is association of PUU virus predominantly with the bank vole, *Clethryonomis glareolus* (sympatric with *Clethryonomis rutilus* and/or with *Clethryonomis rufocanus*) in Europe.[8]

4.2 Problems Associated with Hantavirus Species Definition

Data presented here demonstrate great genetic complexity within the genus *Hantavirus*. Considering the significant geographic variation among hantaviruses harbored by single or several closely related host species, it is not easy to determine whether some recently described virus lineages (for instance, SN-like viruses, or South American hantaviruses associated with *Oligoryzomys* ssp.) constitute true hantavirus species, or whether they should be considered as divergent strains of a single novel virus. In fact, "virus species" is rather a novel concept in virology that has been under development since the early 1990s (see Van Regenmortel et al. 1997, 1999, and references therein). Without attempting to analyze existing literature published on the matter of "virus species", the authors would like instead to touch this topic from the "co-evolutionary" prospective.

In the past, a newly identified arbovirus would be considered a virus serotype (an accepted term for "a distinct virus") if a fourfold or greater two-way difference

[8] To avoid confusion, the authors would like to re-emphasize a clear definition of the "specific rodent host" as "a distinct rodent species/subspecies, which carries a specific hantavirus throughout its entire geographic range". Thus, although many examples exist of the frequently occurring spillover infections in the sympatric "secondary" rodent species, these rodents could not be considered "specific hosts", if observed phenomenon (which could, actually, indicate "host-switching in progress") is limited to the area of sympatricity only.

between this virus and previously recognized related viruses was obtained in virus neutralization assays. However, this is a problem in the case of hantaviruses, which are known to be generally difficult to adapt to cell culture. In addition, a single amino acid substitution in the major neutralizing epitope could potentially allow virus to escape neutralization. Although for obvious biological reasons standard species definitions routinely applied to the multicellular organisms cannot be easily adopted for viruses, advances in the hantavirus molecular genetic and phylogenetic analysis, as well as unique features of the co-evolving hantavirus/host system, prompted virologists to take a new look at the subject. Recently, attempts have been made to introduce a few formal criteria to define hantavirus species (ELLIOTT et al. 1999). These included: (a) a unique ecological niche for each virus species (i.e., a clear association of a new hantavirus with a different primary rodent reservoir species or subspecies); (b) at least 7% amino acid sequence difference from previously characterized hantaviruses (on comparison of the complete glycoprotein precursor and nucleocapsid protein sequences); (c) at least a fourfold difference in two-way cross neutralization tests; and (d) absence of genetic reassortment in nature.

The authors believe that the way to further develop species definition for hantaviruses lays in adapting more widely used general criteria for the definition of biologic species. In MONROE et al. (1999) we suggested a generalized biological definition of species as "the lowest taxonomic unit that is geographically and ecologically contained, genetically (reproductively) isolated, phylogenetically distinct, and self-sufficient". From this prospective, examining known hantavirus/rodent host pairs leads one to the conclusion that all virus strains harbored by a single rodent species are supposed to comprise a single virus species. However, analysis of the mammalogical literature (for instance, HOGAN et al. 1993; for review see CARLETON 1989) shows that in some cases such a definition is a priori limited due to uncertainties in rodent taxonomy. For instance, both the "grassland" form of the deer mouse carrying SN virus and the "forest" form associated with MON virus are currently considered a single rodent species by mammalogists. However, our investigation (MORZUNOV et al. 1996) showed that these two forms of *P. maniculatus* are as distinct in respect to their mitochondrial DNA, as the other group of "deer mice" from Pacific Northwest, which were recently erected in a separate species rank (HOGAN et al. 1993). One way to deal with such uncertainties is to adopt the definition of a "distinct rodent host" as "a reproductively isolated rodent species, or ecologically and/or geographically separated subspecies, which carries a phylogenetically distinct hantavirus, and which is capable of being the sole carrier of this virus in a rodent community". Accepting such a definition, one should recognize *each* of the known SN-like viruses, PH-like viruses, and *Oligoryzomys*-borne South American viruses as a taxonomically distinct entity, i.e., separate virus species. However, although natural and personally favored by the authors, this approach could cause much confusion due to proposed recognition of numerous closely related virus species. An alternative approach could be to lump together as a single virus species all virus strains harbored by the entire rodent genus. Despite its apparent simplicity, such hantavirus systematics would nevertheless be vulnerable to critique. First, it lacks a perfect biological sense inherent to the previously described classification, and, therefore, it could potentially lead to the gross mistake of lumping together very distinct hantaviruses in the cases of host-switching. Second, uncertainties in rodent taxonomy are not limited to the species level. In fact, there are many examples of recent re-classification (usually separation) of existing genera and even higher rodent taxa. Thus, it does not seem possible to avoid making arbitrary decisions, even if this simplistic classification is adopted. Further study, which could include new neutralization data, more detailed identification of the virus-host relationships, more complete genetic characterization of both viruses and their hosts, and, possibly, experimental data on the viability of genetic reassortants between hantavirus strains, will be required to make a final decision on the matter of "hantavirus species".

5 Concluding Remarks

Taken together, current data suggest that both co-speciation of hantaviruses with their specific rodent hosts and biogeographic factors (such as allopatric migrations, geographic separation, and isolation of rodent host populations) have played important roles in establishment of the current genetic diversity and geographic distribution of hantaviruses. Patterns of the hantavirus/rodent host association seen in Fig. 1, suggest that the hantaviruses became associated with the ancestral rodents of the family Muridae before the family division into the subfamilies Murinae, Arvicolinae, and Sigmodontinae around 30 million years ago. Co-speciation appears to be the predominant pattern in the long-term evolution of hantaviruses, although in the process of the host speciation virus could often become extinct, or host-switching events could occur when potential hosts (usually, closely related rodent species) were sympatric.

In conclusion, we should stress that co-evolution of hantaviruses with their rodent hosts is not only of "academic" interest. In fact, knowledge of the phenomena and processes mentioned above can address several questions stated earlier (see Sect. 1). In particular, it improves our ability to predict where new hantaviruses may be found, whether or not they are pathogenic for humans, what are the most appropriate reagents (i.e., reference sera, specific PCR primers, etc.) to use in human diagnostics and rodent surveillance in different geographic areas, and which hantaviruses are the most likely candidates for vaccine development.

Acknowledgments. The authors thank Connie Schmaljohn and Stuart Nichol for the opportunity to write this review on co-evolution of hantaviruses and their rodent hosts. The authors are grateful to John Boone, Irina Gavrilovskaya, Heikki Henttonen, Svetlana Khaiboullina, Åke Lundkvist, Stuart Nichol, Joan Rowe, Olli Vapalahti, and Antti Vaheri for helpful discussions. Some aspects of this work were supported by National Institutes of Health Grants AI36418, AI39808, and AI45059 through the University of Nevada-Reno.

References

Arikawa J, Yoshimatsu K, Wang H, Chen H, Ebihara H, Ogino M, Kariwa H, Takashima I (1999) Antigenic and genetic diversities of hantaviruses isolated in China, Korea, and Japan. VP11.10. In: Abstract Book. XI International Congress of Virology, Sydney, p 293

Avšic-Zupanc TS, Xiao SY, Stojanovic R, Gligic A, van der Groen J, Le Duc JW (1992) Characterization of Dobrava virus: a hantavirus from Slovenia, Yugoslavia. J Med Virol 38:132–137

Avšic-Zupanc T, Toney A, Anderson K, Chu YK, Schmaljohn C (1995) Genetic and antigenic properties of Dobrava virus: a unique member of the hantavirus genus, family *Bunyaviridae*. J Gen Virol 76:2801–2808

Avšic-Zupanc T, Petrovec M, Furlan P, Kaps R, Elgh F, Lundkvist Å (1999) Hemorrhagic fever with renal syndrome (HFRS) in the Dolenjska region of Slovenia – a 10-year survey. Clinical Infect Dis 28:860–865

Avšic-Zupanc T, Nemirov K, Petrovec M, Trilar T, Poljak M, Vaheri A, Plyusnin A (2000) Genetic analysis of wild-type Dobrava hantavirus in Slovenia: co-existence of two distinct genetic lineages within the same natural focus. J Gen Virol 81:1747–1755

Baker RH (1968) Habitats and distribution In: King JA (ed) Biology of *Peromyscus* (Rodentia). Spec. Publ Am Soc Mammalogists, Lawrence, Kansas, pp 98–126

Baker RJ, Robbins LW, Stangl Jr FB, Birney EC (1983) Chromosomal evidence for a major subdivision in *Peromyscus leucopus*. J Mammal 64:356–359

Bharadwaj M, Botten J, Torrez-Martinez N, Hjelle B (1997) Rio Mamore virus: genetic characterization of a newly recognized hantavirus of the pygmy rice rat, Oligoryzomys microtis, from Bolivia. Am J Trop Med Hyg 57:368–374

Blair WF (1950) Ecological factors in speciation of *Peromyscus*. Evolution 4:253–275

Bowen MD, Kariwa H, Rollin PE, Peters CJ, Nichol ST (1995) Genetic characterization of a human isolate of Puumala hantavirus from France. Virus Research 38:279–289

Bowen MD, Gelbmann W, Ksiasek T, Nichol ST, Nowotny N (1997) Puumala virus and two genetic variants of Tula virus are present in Austrian rodents. J Med Virol 53:174–181

Brummer-Korvenkontio M, Vaheri A, Hovi T, von Bonsdorff CH, Vuorimies J, Manni T, Penttinen K, Oker-Blom N, Lahdevirta J (1980) Nephropathia epidemica: detection of antigen in bank voles and serologic diagnosis of human infection. J Infect Dis 141:131–134

Brummer-Korvenkontio M, Vapalahti O, Henttonen H, Koskela P, Kuusisto P, Vaheri A (1999) Epidemiological study of Nephropathia epidemica in Finland 1989–1996. Scand J Infect Dis 31:1–9

Carey DE, Reuben R, Panicker KN, Shope RE, Myers RM (1971) Thottapalayam virus: a presumptive arbovirus isolated from a shrew in India. Indian J Med Res 59:1758–1760

Carleton MD (1989) Systematics and evolution. In: Kirkland Jr GL, Layne JN (eds) Advances in the study of *Peromyscus* (Rodentia). Texas Tech University Press, Lubbock, Texas, pp 7–141

Carleton MD, Musser GG (1989) Systematic studies of oryzomyine rodents (Muridae, Sigmodontinae): a synopsis of *Microryzomys*. Bulletin of the American Museum of Natural History 191:1–83

Chelomina GN, Suzuki H, Tsuchiva K, Moriwaki K, Liapunova EA, Vorontsov NN (1998) Sequencing of the mtDNA cytochrome B gene and reconstruction of the matriarchal relationships between wood and field mice of the genus *Apodemus* (*Muridae*, *Rodentia*). Genetika 34:650–661 (in Russian)

Chelomina GN (1998) Molecular phylogeny of forest and field mice of the genus *Apodemus* (*Muridae*, *Rodentia*) based on the data on restriction analysis of total nuclear DNA. Genetika 34:1286–1292 (in Russian)

Chernukha YuG, Evdokimova OA, Chechovich AV (1986) Results of karyologic and immunobiological studies of the striped field mouse (*Apodemus agrarius*) from different areas of its range. Zoological Journal 65:471–475 (in Russian)

Childs JE, Ksiazek TG, Spiropoulou CF, Krebs JW, Morzunov S, Maupin GO, Gage KL, Rollin PE, Sarisky J, Enscore RE, Frey JK, Peters CJ, Nichol ST (1994) Serologic and genetic identification of *Peromyscus maniculatus* as the primary rodent reservoir for a new hantavirus in the southwestern United States. J Infect Dis 169:1271–1280

Chizhikov VE, Spiropoulou SF, Morzunov SP, Monroe MC, Peters CJ, Nichol ST (1995) Complete genetic characterization and analysis of isolation of Sin Nombre virus. J Virol 69:8132–8136

Dickerman AW, Yates TL (1995) Systematics of *Oligoryzomys*: protein-electrophoretic analyses. J Mammal 76:172–188

Duchin JS, Koster FT, Peters CJ, Simpson GL, Tempest B, Zaki S, Ksiazek TG, Rollin PE, Nichol ST, Umland ET, Moolenaar RL, Reef SE, Nolte KB, Gallaher MM, Butler JC, Breman RF, The

Hantavirus Study Group (1994) Hantavirus Pulmonary Syndrome: a clinical description of 17 patients with a newly recognized disease. New England J Med 330:949–955

Dzagurova T, Tkachenko E, Slonova R, Ivanov L, Ivanidze E, Markeshin S, Dekonenko A, Niklasson B, Lundkvist Å (1995) Antigenic relationships of hantavirus strains analysed by monoclonal antibodies. Arch Virol 140:1763–1773

Elliott LH, Ksiazek TG, Rollin PE, Spiropoulou CF, Morzunov S, Monroe M, Goldsmith CS, Humphrey CD, Zaki SR, Krebs JW, Maupin G, Gage K, Childs JE, Nichol ST, Peters CJ (1994) Isolation of the causative agent of hantavirus pulmonary syndrome. Am J Trop Med Hyg 51:102–108

Elliott RM, Bouloy M, Calisher CH, Goldbach R, Moyer JT, Nichol ST, Pettersson R, Plyusnin A, Schmaljohn CS (1999) Bunyaviridae. In: van Regenmortel MHV, Fauquet CM, Bishop DHL, Carsten EB, Estes MK, Lemon SM, Maniloff J, Mayo MA, McGeoch DJ, Pringle CR, Wickner RB (eds) Virus taxonomy: VIIth report of the International Committee on Taxonomy of Viruses. Academic Press, San Diego, California, pp 599–631

Engel SR, Hogan KM, Taylor JF, Davis SK (1998) Molecular systematics and paleobiology of the South American sigmodontine rodents. Molec Biol Evol 15:35–49

Escutenaire S, Chalon P, Heyman P, Van der Auwer G, Van der Groen G, Verhagen R, Thomas I, Karelle-Bui L, Vaheri A, Pastoret P-P, Plyusnin A (2000) Genetic characterization of Puumala hantavirus strains from Belgium: evidence for a distinct phylogenetic lineage. Virus Res, in press

Fedorov V, Goropashnaya A, Jarrell GH, Fredga K (1999) Phylogeographic structure and mitochondrial DNA variation in true lemmings (*Lemmus*) from the Eurasian Arctic. Biol J Linnean Soc 66:357–371

Fulhorst CF, Monroe MC, Salas RA, Duno G, Utrera A, Ksiazek TG, Nichol ST, de Manzione NM, Tovar D, Tesh RB (1997) Isolation, characterization and geographic distribution of Caño Delgadito virus, a newly discovered South American hantavirus (family Bunyaviridae). Virus Res 51:159–171

Gavrilovskaya I, Boiko V (1986) Hemorrhagic fever with renal syndrome. Scientific Reviews of The Institute for Medical Information (VNIIMI), Moscow, pp 1–74 (in Russian)

Giebel LB, Stohwasser R, Zöller L, Bautz EKF, Darai G (1989) Determination of the coding capacity of the M genome segment of Nephropathia epidemica virus strain Hällnäs B1 by molecular cloning and nucleotide sequence analysis. Virology 172:498–505

Golovljova I, Vasilenko V, Prükk T, Sjölander BK, Plyusnin A, Lundkvist Å (2000) Dobrava and Puumala hantaviruses cause HFRS in Estonia. Eur J Clin Microbiol Infect Dis, in press

Hall ER (1981) The Mammals of North America (2nd ed.). John Wiley and Sons, New York, 2:601–1181

Henderson WW, Monroe MC, St Jeor SC, Thayer WP, Rowe JE, Peters CJ, Nichol ST (1996) Naturally occurring Sin Nombre virus genetic reassortants. Virology 214:602–610

Henttonen H, Kaikusalo A (1993) Lemming movements. In: Stenseth NC, I. R.A. (eds) The biology of lemmings. Academic Press, pp 157–186

Hershkovitz P (1966) Mice, land bridges and Latin American faunal interchange. In: Wenzel RL, Tipton VJ (eds) Ectoparasites of Panama. Field Museum of Natural History, Chicago, pp 725–751

Hjelle B, Jenison S, Torrez-Martinez N, Yamada T, Nolte K, Zumwalt R, MacInnes K, Myers G (1994a) A novel hantavirus associated with an outbreak of fatal respiratory disease in the southwestern United States: evolutionary relationships to known hantaviruses. J Virol 68:592–596

Hjelle B, Chavez-Giles F, Torrez-Martinez N, Yates T, Sarisky J, Webb J, Ascher M (1994b) Genetic identification of a novel hantavirus of the harvest mouse *Reithrodontomys megalotis*. J Virol 68: 6751–6754

Hjelle B, Chavez-Giles F, Torrez-Martinez N, Yamada T, Sarisky J, Ascher M, Jenison S (1994c) Dominant glycoprotein epitope of four corners hantavirus is conserved across a wide geographical area. J Gen Virol 75:2881–2888

Hjelle B, Anderson B, Torrez-Martinez N, Song W, Gannon WL, Yates TL (1995a) Prevalence and geographic genetic variation of hantaviruses of New World harvest mice (*Reithrodontomys*): identification of a divergent genotype from a Costa Rican *Reithrodontomys mexicanus*. Virology 207: 452–459

Hjelle B, Krolikowski J, Torrez-Martinez N, Chavez-Giles F, Vanner C, Laposata E (1995b) Phylogenetically distinct hantavirus implicated in a case of hantavirus pulmonary syndrome in the Northeastern United States. J Med Virol 46:21–27

Hjelle B, Lee SW, Song W, Torrez-Martinez N, Song JW, Yanagihara R, Gavrilovskaya I, Mackow ER (1995c) Molecular linkage of hantavirus pulmonary syndrome to the white-footed mouse, *Peromyscus leucopus*: genetic characterization of the M genome of New York virus. J Virol 69:8137–8141

Hogan KM, Hedin MC, Koh HS, Davis SK, Greenbaum IF (1993) Systematic and taxonomic implications of karyotypic, electrophoretic, and mitochondrial-DNA variation in *Peromyscus* from the Pacific Northwest. J Mammal 74:819–831

Hooper ET (1968) Classification. In: King JA (ed) Biology of *Peromyscus*. Spec Publ Am Soc Mammalogists 2, Lawrence, Kansas, pp 27–74

Hörling J, Lundkvist Å, Plyusnin A, Jaarola M, Tegelström H, Persson K, Lehväslahio H, Hörnfeldt B, Vaheri A, Niklasson B (1996a) Distribution and genetic heterogeneity of Puumala virus in Sweden. J Gen Virol 77:2555–2562

Hörling J, Chizhikov V, Lundkvist Å, Jonsson M, Ivanov L, Dekonenko A, Niklasson B, Dzagurova T, Peters CJ, Tkachenko E, Nichol S (1996b) Khabarovsk virus: a phylogenetically and serologically distinct hantavirus isolated from *Microtus fortis* trapped in far-east Russia. J Gen Virol 77:687–694

Jaarola M, Tegelström H, Fredga K (1999) Colonization history in Fennoscandian rodents. In: Racey PA, Bacon PJ, Dallas JF, Piertney SB (eds) Molecular genetics in animal ecology. Biol J Linnean Soc 68:113–127

Johnson AM, Bowen MD, Ksiazek TG, Williams RJ, Bryan RT, Mills JN, Peters CJ, Nichol ST (1997) Laguna Negra virus associated with HPS in western Paraguay and Bolivia. Virology 238:115–127

Kariwa H, Isegawa Y, Arikawa J, Takashima I, Ueda S, Yamanishi K, Hashimoto N (1994) Comparison of nucleotide sequences of M genome segments among Seoul virus strains isolated from Eastern Asia. Virus Res 33:27–38

Kariwa H, Yoshimatsu K, Sawabe J, Yokota E, Arikawa J, Takashima I, Fukushima H, Lundkvist Å, Shubin, FN, Isachkova, LM, Slonova RA, Leonova GN, Hashimoto N (1999) Genetic diversities of hantaviruses among rodents in Hokkaido, Japan and Far East Russia. Virus Res 59:219–228

Kariwa H, Yoshimatsu K, Araki K, Murphy ME, Ebihara H, Ogino M, Mizutani T, Arikawa J, Chayama K, Kumada H, Takashima I (2000) Detection of hantavirus antibodies among patients with hepatitis of unknown etiology in Japan. Microbiol Immunol, in press

Khaiboullina SF, St. Jeor SC, Morzunov SP (1997) Genetic variability and natural reassortment of the Puumala hantavirus in Bashkiria, Russia. S:210. In: Emergence and re-emergence of negative strand viruses. 10th International Conference on Negative Strand Viruses, Dublin, Ireland, p 155

Khan AS, Spiropoulou SF, Morzunov S, Zaki SR, Kohn MA, Nawas SR, McFarland L, Nichol ST (1995) Fatal illness associated with a new hantavirus in Louisiana. J Med Virol 46:281–286

Kratochvil J (1959) Hrabos polni, *Microtus arvalis*. Czechoslovak Academy of Sciences Publ., Prague, p 359

Ksiazek TG, Nichol ST, Mills JN, Groves MG, Wozniak A, McAdams S, Monroe MC, Johnson AM, Martin ML, Peters CJ, Rollin PE (1997) Isolation, genetic diversity, and geographic distribution of Bayou virus (Bunyaviridae: hantavirus). Am J Trop Med Hyg 57:445–448

Lee HW, Lee PW (1978) Isolation of the etiologic agent of Korean hemorrhagic fever. J Infect Dis 137:289–308

Lee HW, Lee PW, Baek LJ, Chu YK (1990) Geographical distribution of hemorrhagic fever with renal syndrome and hantaviruses. Archives of Virology (Supplement) 1:5–18

Levis S, Morzunov SP, Rowe JE, Enria DA, Pini N, Calderon G, Sabattini M, St. Jeor SC (1998) Genetic diversity and epidemiology of hantaviruses in Argentina. J Infect Dis 177:529–538

Liang M, Li D, Xiao S-H, Hang C, Rossi CA, Schmaljohn CS (1994) Antigenic and molecular characterization of hantavirus isolates from China. Virus Res 31:219–233

Lopez N, Padula P, Rossi C, Lazaro ME, Franze-Fernandez MT (1996) Genetic identification of a new hantavirus causing severe pulmonary syndrome in Argentina. Virology 220:223–226

Lopez N, Padula P, Rossi C, Miguel S, Edelstein A, Ramirez E, Franze-Fernandez MT (1997) Genetic characterization and phylogeny of Andes virus and variants from Argentina and Chile. Virus Res 50:77–84

Lundkvist Å, Apekina N, Myasnikov Y, Vapalahti O, Vaheri A, Plyusnin A (1997) Dobrava hantavirus outbreak in Russia. Lancet 350:781–782

Lundkvist Å, Vasilenko V, Golovljova I, Plyusnin A, Vaheri A (1998a) Human Dobrava hantavirus infections in Estonia. Lancet 352:369

Lundkvist Å, Wiger D, Hörling J, Sjölander KB, Plyusnina A, Mehl R, Vaheri A, Plyusnin A (1998b) Isolation and characterization of Puumala hantavirus from Norway: evidence for a distinct phylogenetic sublineage. J Gen Virol 78:2603–2614

Marshall LG (1979) A model for paleobiogeography of South American cricetine rodents. Paleobiology 5:126–132

Mitchell-Jones AJ, Amori G, Bogdanowicz W, Krystufek B, Reijnders PJH, Spitzenberger F, Stubbe M, Thissen JBM, Vohralik V, Zima J (1999) The atlas of European mammals. London, T and AD Poyser Natural History, p 484

Monroe MC, Morzunov SP, Johnson AM, Bowen MD, Artsob H, Yates T, Peters CJ, Rollin PE, Ksiazek TG, Nichol ST (1999) Genetic diversity and distribution of *Peromyscus*-borne hantaviruses in North America and comparison with other hantaviruses. Emerging Infect Dis 5:75–86

Morzunov SP, Feldmann H, Spiropoulou CF, Semenova VA, Rollin PE, KsiazekTG, Peters CJ, Nichol ST (1995) A newly recognized virus associated with a fatal case of hantavirus pulmonary syndrome in Louisiana. J Virol 69:1980–1983

Morzunov SP, St Jeor SC, Yates TL, Rowe J, Ksiazek TG, Peters CJ, Nichol ST (1996) Co-evolution of North American *Peromyscus*-borne hantaviruses with specific rodent hosts. W11-5. In: Abstract book. Xth International Congress of Virology, Jerusalem, Israel, p 23

Morzunov SP, Rowe JE, Monroe MC, Ksiazek TG, Peters CJ, St. Jeor SC, Nichol ST (1998) Genetic analysis of the diversity and origin of hantaviruses in *Peromyscus leucopus* mice in North America. J Virol 72:57–64

Morzunov SP, Khaiboullina SF, St. Jeor SC, Khasanova SS, Lundkvist Å, Plyusnin A (1999) Co-evolution of distinct lineages of the Puumala virus with phylogenetically distinct forms of the bank vole. VP11.15. In: Abstract Book. XIth International Congress of Virology, Sydney, p 295

Musser GG, Carleton MD (1993) Family Muridae. In: Wilson DE, Reader DM (eds) Mammal species of the world. Smithsonian Institution Press, New York, pp 501–755

Nelson K, Baker RJ, Honeycutt RL (1987) Mitochondrial DNA and protein differentiation between hybridizing cytotypes of the white-footed mouse, *Peromyscus leucopus*. Evolution 4:864–872

Nemirov K, Vapalahti O, Lundkvist Å, Vasilenko V, Golovjova I, Plyusnina A, Niemiaa J, Laakkonen J, Henttonen H, Vaheri A, Plyusnin A (1999) Isolation and characterization of Dobrava hantavirus carried by the striped field mouse (*Apodemus agrarius*) in Estonia. J Gen Virol 80:371–379

Nichol ST, Spiropoulou CF, Morzunov SP, Rollin PE, Ksiazek TG, Feldmann H, Sanchez A, Childs J, Zaki S, Peters CJ (1993) Genetic identification of a hantavirus associated with an outbreak of acute respiratory illness. Science 262:914–917

Nichol ST, Ksiazek TG, Rollin PE, Peters CJ (1996) Hantavirus pulmonary syndrome and newly described hantaviruses in the United States. In: Elliott RM (ed) The Bunyaviridae. Plenum Press, New York, pp 269–280

Nichol ST (1999) Genetic analysis of hantaviruses and their host relationships. In: Saluzzo JF, Dodet B (eds) Emergence and control of rodent-borne viral diseases. Elsevier Publishers, Paris, pp 99–109

Osgood WH (1909) Revision of the mice of the American genus *Peromyscus*. N Amer Fauna 28:1–285

Padula PJ, Edelstein A, Miguel SD, Lopez NM, Rossi CM, Rabinovich RD (1998) Hantavirus pulmonary syndrome outbreak in Argentina: molecular evidence for person-to-person transmission of Andes virus. Virology 241(2):323–330

Padula P, Miguel S, Martinez V, Edelstein A, Colavecchia S, Gonzalez DVM, Rabinovich R, Segura E (1999) Human respiratory hantavirus infections in southern cone of America. VP11.11. In: Abstract Book. XIth International Congress of Virology, Sydney, p 294

Papa A, Johnson AM, Stockton PC, Bowen MD, Spiropoulou CF, Alexiou-Daniel S, Ksiazek TG, Nichol ST, Antoniadis A (1998) Retrospective serological and genetic study of the distribution of hantaviruses in Greece. J Med Virol 55:321–327

Parrington MA, Kang CY (1990) Nucleotide sequence analysis of the S genomic segment of Prospect Hill virus: comparison with the prototype hantavirus. Virology 175:167–175

Parrington MA, Lee PW, Kang CY (1991) Molecular characterization of the Prospect Hill virus M RNA segment: a comparison with the M RNA segments of other hantaviruses. J Gen Virol 72:1845–1854

Pilaski J, Feldmann H, Morzunov S, Rollin PE, Ruo SL, Lauer B, Peters CJ, Nichol S (1994) Genetic identification of a new Puumala virus strain causing severe hemorrhagic fever with renal syndrome in Germany. J Infect Dis 170:1456–1462

Pilaski J, Ivanov A, Göbels K, Lewandowski B, Kupfer M, Feldmann H (1999) Different hantaviruses are circulating in Germany. V11RS. 2. In: Abstract Book. XIth International Congress of Virology, Sydney, p 291

Plyusnin A, Vapalahti O, Ulfves K, Lehväslaiho H, Apekina N, Gavrilovskaya I, Blinov V, Vaheri A (1994a) Sequences of wild Puumala virus genes show a correlation of genetic variation with geographic origin of the strains. J Gen Virol 75:405–409

Plyusnin A, Vapalahti O, Lankinen H, Lehväslaiho H, Apekina N, Myasnikov Y, Kallio-Kokko H, Henttonen H, Lundkvist Å, Brummer-Korvenkontio M, Gavrilovskaya I, Vaheri A (1994b) Tula virus: a new hantavirus carried by European common voles (*Microtus arvalis* and *Microtus rossiaemeridionalis*). J Virol 68:7833–7839

Plyusnin A, Vapalahti O, Lehväslaiho H, Apekina N, Mikhailova T, Gavrilovskaya I, Laakonen J, Niemimaa J, Henttonen H, Brummer-Korvenkontio M, Vaheri A (1995a) Genetic variation of wild Puumala viruses within the serotype, local rodent population and individual animal. Virus Res 38:25–41

Plyusnin A, Cheng Y, Vapalahti O, Pejcoch M, Unar J, Jelinkova Z, Lehväslaiho H, Lundkvist Å, Vaheri A (1995b) Genetic variation in Tula hantaviruses: sequence analysis of the S and M segments of strains from Central Europe. Virus Res 39:237–250

Plyusnin A, Vapalahti O, Vaheri A (1996a) Hantaviruses: genome structure, expression and evolution. J Gen Virol 77:2677–2687

Plyusnin A, Vapalahti O, Lundkvist Å, Henttonen H, Vaheri A (1996b) Newly recognized hantavirus in Siberian lemmings. Lancet 347:1835–1836

Plyusnin A, Hörling J, Kanerva M, Mustonen J, Cheng Y, Partanen J, Vapalahti O, Kukkonen SKJ, Niemimaa J, Henttonen H, Niklasson B, Lundkvist Å, Vaheri A (1997a) Puumala hantavirus genome in patients with nephropathia epidemica: correlation of PCR positivity with HLA haplotype and link to viral sequences in local rodents. J Clinical Microbiology 35:1090–1096

Plyusnin A, Vapalahti O, Vasilenko V, Henttonen H, Vaheri A (1997b) Dobrava hantavirus in Estonia: does the virus exist throughout Europe? Lancet 349:1369–1370

Plyusnin A, Nemirov K, Apekina N, Plyusnina A, Lundkvist Å, Vaheri A (1999a) Dobrava hantavirus in Russia. Lancet 353:207

Plyusnin A, Mustonen J, Asikainen K, Plyusnina A, Niemimaa J, Henttonen H, Vaheri A (1999b) Analysis of Puumala hantavirus genome in patients with nephropathia epidemica and rodent carriers from the sites of infection. J Med Virol 59:397–405

Powers AM, Mercer DR, Watts DM, Guzman H, Fulhorst CF, Popov VL, Tesh RB (1999) Isolation and genetic characterization of a hantavirus (*Bunyaviridae*: Hantavirus) from a rodent, *Oligoryzomys microtis* (Muridae), collected in northeastern Peru. Am J Trop Med Hyg 61:92–98

Ravkov EV, Rollin PE, Ksiazek TG, Peters CJ, Nichol ST (1995) Genetic and serologic analysis of the Black Creek Canal virus and its association with human disease and *Sigmodon hispidus* infection. Virol 210:482–489

Rawlings JA, Torrez-Martinez N, Neill SU, Moore GM, Hicks BN, Pichuantes S, Nguyen A, Bharadwaj M, Hjelle B (1996) Cocirculation of multiple hantaviruses in Texas, with characterization of the small (S) genome of a previously undescribed virus of cotton rats (*Sigmodon hispidus*). Am J Trop Med Hyg 55:672–679

Rollin PE, Ksiazek TG, Elliott LH, Ravkov EV, Martin ML, Morzunov S, Livingstone W, Monroe M, Glass G, Ruo S, Khan AS, Childs JE, Nichol ST, Peters CJ (1995) Isolation of Black Creek Canal virus, a new hantavirus from *Sigmodon hispidus* in Florida. J Med Virol 46:35–39

Rowe JE, St. Jeor SC, Riolo J, Otteson EW, Monroe MC, Henderson WW, Ksiazek TG, Rollin PE, Nichol ST (1995) Coexistence of several novel hantaviruses in rodents indigenous to North America. Virology 213:122–130

Schmaljohn CS, Arikawa J, Hasty SE, Rasmussen L, Lee HW, Lee PW, Dalrimple JM (1988) Conservation of antigenic properties and sequences encoding the envelope proteins of prototype Hantaan virus and two virus isolates from Korean haemorrhagic fever patients. J Gen Virol 69:1949–1955

Schmaljohn AL, Li D, Negley DL, Bressler DS, Turell MJ, Korch GW, Ascher MS, Schmaljohn CS (1995) Isolation and initial characterization of a newfound hantavirus from California. Virol 206:963–72

Schmaljohn CS, Hjelle B (1997) Hantaviruses: a global disease problem. Emerging Infect Dis 3:95–104

Sibold C, Sparr S, Schulz A, Labuda M, Kozuch O, Lysy J, Krüger DH, Meisel H (1995) Genetic characterization of a new hantavirus detected in *Microtus arvalis* from Slovakia. Virus Genes 10:277–281

Sibold C, Meisel H, Kruger DH, Labuda M, Lysy J, Kozuch O, Pejcoch M, Vaheri A, Plyusnin A (1999a) Recombination in Tula hantavirus evolution: analysis of genetic lineages from Slovakia. J Virol 73:667–675

Sibold C, Meisel H, Lundkvist A, Schulz A, Cifire F, Ulrich R, Kozuch O, Labuda M, Kruger DH (1999b) Short report: simultaneous occurrence of Dobrava, Puumala, and Tula Hantaviruses in Slovakia. Amer J Tropical Med Hygiene 61:409–411

Simpson GG (1969) South American mammals. In: Fittkau EJ, Illies J, Klinge H, Schwabe GH, Sioli H (eds) Biogeography and ecology in South America. W.N.V. Publishers, The Hague, pp 879–909

Smith MF, Patton JL (1993) The diversification of South American murid rodents: evidence from mitochondrial DNA sequence data for the akodontine tribe. Biological Journal of the Linnean Society 50:149–177

Song W, Torrez-Martinez N, Irwin W, Harrison FJ, Davis R, Ascher M, Jay M, Hjelle B (1995) Isla Vista virus: a genetically novel hantavirus of the California vole *Microtus californicus*. J General Virol 76:3195–3199

Song JW, Baek LJ, Nagle JW, Schlitter D, Yanagihara R (1996) Genetic and phylogenetic analyses of hantaviral sequences amplified from archival tissues of deer mouse (*Peromyscus maniculatus nubiterrae*) captured in the eastern United States. Arch Virol 141:959–967

Song JW, Lee YJ, Baek LJ, Song KJ, Skrok A, Markowski J, Bratosievicz J, Kordek R, Libersky PP, Yanagihara R (1998) Characterization of Tula Hantavirus isolated from *Microtus arvalis* captured in Poland. In: Abstract book of the Fouth International Conference on HFRS and Hantaviruses, Atlanta, p 78

Song JW, Baek LJ, Kim SH, Lee YJ, Yanagihara R, Song KJ (1999) Muju virus: a Puumala-related hantavirus harboured by *Eothenomys regulus* in Korea. V11.RS. 1. In: Abstract Book. XIth International Congress of Virology, Sydney, p 290

Spiropoulou CF, Morzunov S, Feldmann H, Sanchez A, Peters CJ, Nichol ST (1994) Genome structure and variability of a virus causing hantavirus pulmonary syndrome. Virology 200:715–723

Stangl Jr FB (1986) Aspects of a contact zone between two chromosomal races of *Peromyscus leucopus* (Rodentia: Cricetidae). J Mammal 67:465–473

Stohwasser R, Giebel, LB, Zöller L, Bautz EKF, Darai G (1990) Molecular characterization of the RNA S segment of nephropathia epidemica virus strain Hällnäs B1. Virology 174:79–86

Tegelström H (1987) Transfer of mitochondrial DNA from the Northern red-backed vole (*Clethrionomys rutilus*) to the bank vole (*C. glareolus*). J Mol Evol 24:218–227

Tegelström H, Wyöni PI, Gelter H, Jaarola M (1988) Concordant divergence in protein and mitochondrial DNA berween two vole species in the genus *Clethrionomys*. Biochem Genetics 26:223–237

Toro J, Vega JD, Khan AS, Mills JN, Padula P, Terry W, Yadon Z, Valderrama R, Ellis BA, Pavletic C, Cerda R, Zaki S, Shieh WJ, Meyer R, Tapia M, Mansilla C, Baro M, Vergara JA, Concha M, Calderon G, Enria D, Peters CJ, Ksiazek TG (1998) An outbreak of hantavirus pulmonary syndrome, Chile, 1997. Emerg Infect Dis 4:687–694

Torrez-Martinez N, Bharadwaj M, Goade D, Delury J, Moran P, Hicks B, Nix B, Davis JL, Hjelle B (1998) Bayou virus-associated hantavirus pulmonary syndrome in Eastern Texas: identification of the rice rat, *Oryzomys palustris*, as reservoir host. Emerg Infect Dis 4:105–111

Van Den Bussche RA, Chesser RA, Hamilton MJ, Bradley RD, Porter CA, Baker RJ (1993) Maintenance of a narrow hybrid zone in *Peromyscus leucopus*: a test of alternative models. J Mammal 74:832–845

Van Regenmortel MH, Bishop DH, Fauquet CM, Mayo MA, Maniloff J, Calisher CH (1997) Guidelines to the demarcation of virus species. Arch Virol 142:1505–1518

Van Regenmortel MH (1999) How to write the names of virus species. Arch Virol 144:1041–1042

Vapalahti O, Kallio-Kokko H, Salonen EM, Brummer-Korvenkontio M, Vaheri A (1992) Cloning and sequencing of Puumala virus Sotkamo strain S and M RNA segments: evidence for strain variation in hantaviruses and expression of the nucleocapsid protein. J Gen Virol 73:829–838

Vapalahti O, Lundkvist Å, Kukkonen SKJ, Cheng Y, Gilljam M, Kanerva M, Manni T, Pejcoch M, Niemimaa J, Kaikusalo A, Henttonen H, Vaheri A, Plyusnin A (1996) Isolation and characterization of Tula virus, a distinct serotype in the genus Hantavirus, family *Bunyaviridae*. J Gen Virol 77: 3063–3067

Vapalahti O, Lundkvist Å, Fedorov V, Conroy C, Hirvonen S, Plyusnina A, Nemirov K, Fredga K, Cook J, Niemimaa J, Kaikusalo A, Henttonen H, Vaheri A, Plyusnin A (1999) Isolation and characterization of a hantavirus from *Lemmus sibiricus*: evidence for host-switch during hantavirus evolution. J Virol 73:5586–5592

Waters JH (1963) Biochemical relationships of the mouse *Peromyscus* in New England. Systematic Zoology 12:122–132

Webb SD (1976) Mammalian faunal dynamics of the Great American Interchange. Paleobiology 2: 220–234

Williams RJ, Bryan RT, Mills JN, Palma RE, Vera I, de Velasquez F, Baez E, Schmidt WE, Figueroa RE, Peters CJ, Zaki SR, Khan AS, Ksiazek TG (1997) An outbreak of hantavirus pulmonary syndrome in western Paraguay. Am J Trop Med Hyg 57:274–282

Wolff JO (1986) Getting along in Appalachia. Natural History 9:45–48

Wolff JO (1989) Social behavior In: Kirkland Jr GL, Layne JN (eds) Advances in the study of *Peromyscus* (Rodentia). Texas Tech University Press, Lubbock, Texas, pp 271–291

Xiao SY, Liang M, Schmaljohn CS (1993a) Molecular and antigenic characterization of HV114, a hantavirus isolated from a patient with haemorrhagic fever with renal syndrome in China. J Gen Virol 74:1657–1659

Xiao SY, Spik KW, Li D, Schmaljohn CS (1993b) Nucleotide and deduced amino acid sequences of the M and S genome segments of two Puumala virus isolates from Russia. Virus Res 30:97–103

Xiao SY, Leduc JW, Chu YK, Schmaljohn CS (1994) Phylogenetic analyses of virus isolates in the genus *Hantavirus*, family *Bunyaviridae*. virol 198:205–217

Modeling Hantavirus Maintenance and Transmission in Rodent Communities

B. HJELLE[1] and T. YATES[2]

1 Hantavirus-Rodent Co-Evolution

Analysis of the phylogenetic relationships among hantaviruses suggests that hantaviruses have had a longstanding co-evolutionary history with their predominant rodent carriers (SCHMALJOHN and HJELLE 1997; Nichol, in press). Closely related hantaviruses are generally derived from rodents that are themselves closely related, while deeper evolutionary branches in the virus phylogenetic tree are associated with deeper branches in the rodent phylogenetic tree. As a first approximation, hantavirus evolution resembles that of an extrachromosomal genetic element of murid rodents more than it does that of an autonomous, horizontally transmitted agent. Those hantaviruses that have a clearly identified reservoir host are associated with rodents of the family Muridae, subfamilies Murinae, Arvicolinae, and Sigmodontinae. A single hantavirus isolate, Thottapalayam, was obtained from a non-murid host, the shrew *Suncus murinus* (CAREY et al. 1971). It is unclear whether insectivores are truly reservoirs for hantaviruses, or rare incidental hosts. If

[1] Departments of Pathology, Biology and Molecular Genetics and Microbiology University of New Mexico, Albuquerque, NM 87131, USA
e-mail: bhjelle@salud.unm.edu
[2] Department Biology, University of New Mexico, Albuquerque, NM 87131, USA
e-mail: tyates@sevilleta.unm.edu

the former were true then these viruses would most likely represent ancient relatives of those found in murid rodents, given the temporal depth of the divergence between Insectivora and Rodentia. It appears to us to be somewhat more likely that Thottapalayam represents a recent host-switching event, since even the high divergence of that virus from other hantaviruses is not as large as is the divergence between Insectivora and Rodentia.

Rodents of the family Muridae, and presumably their associated hantaviruses, appear to have had most of their evolutionary history restricted to the Old World, with specific virus clades evolving within rodent species of the murid subfamilies Murinae and Arvicolinae. Members of the family Muridae are believed to have arrived in the New World via the Bering land bridge approximately 20–30 million years before present.

In the New World, no members of the subfamily Murinae were present until recent times when western travelers introduced the commensal rats *Rattus norvegicus* and *R. rattus* and the house mouse *Mus musculus*. However, rodents of the subfamily Arvicolinae are indigenous to both the Americas and to Eurasia. Despite the considerable geographic barriers that have historically segregated the biota of the eastern and western hemispheres, arvicoline rodents from both Eurasia and North America harbor rather closely related hantaviruses. In phylogenetic analyses, New World arvicoline-borne hantaviruses, although distinct from Old World forms, are clearly a member of the same clade as Old World Arvicolinae and thus shared a more recent common ancestor with members of that subfamily than with species in other murid subfamilies. It is difficult to establish precisely when and where these vole-borne hantaviruses first evolved, when they crossed over the Bering land bridge, and whether additional exchange of rodents and viruses occurred after their original migration. As a first approximation, the most important factor determining the genotypic characteristic of each vole-borne hantavirus is the rodent species with which it is associated, not its geographic placement.

Rodents of the subfamily Sigmodontinae arose in North America. Both North and South American members of this subfamily appear to have evolved from the late Miocene genus, *Copemys*, which was the only sigmodontine known from North America from 16–9 million years ago (MYA) (Baskin 1989). A new clade of hantaviruses arose along with the new rodent clade, probably from an ancestor that is shared with those of the arvicoline-borne hantaviruses. A number of distinct North American hantaviruses have been identified among sigmodontine rodents, including *Peromyscus maniculatus* (Sin Nombre, or SN virus); *P. leucopus* (New York virus); *Reithrodontomys megalotis* (El Moro Canyon virus); *R. mexicanus* (Rio Segundo virus); *Sigmodon* spp. (Muleshoe virus; Black Creek Canal virus); and *Oryzomys palustris* (Bayou virus) (Schmaljohn and Hjelle 1997; Mertz et al. 1997 and references therein).

The Bolivar Trench, a body of water that has separated North and South America since the Permian, was breached by a land bridge approximately 7 MYA. Sigmodontine rodents invaded South America at or before this time and underwent a rapid adaptive radiation. The occurrence of two modern genera, *Auliscomys* and *Bolomys*, in Argentina 3 MYA (Reig 1978) is also compatible with the possibility that sigmodontines may have arrived in South America at an earlier date by waif dispersal. Currently, there are at least 300 species in 62 genera of South American sigmodontines. The occurrence of two South American genera, *Sigmodon* and *Oryzomys*, in North America suggests a later re-invasion from the south.

Interestingly, the hantaviruses thus identified in South American sigmodontine rodents are considerably less genetically complex than those found in North America. In one of the clearest examples of host switching by hantaviruses, the Laguna Negra (LN) virus of Paraguay, apparently was transferred from an oryzomine species to a phyllotine species (Johnson et al. 1997). Phylogenetically, LN virus is closely allied with the oryzomine-borne hantaviruses Andes virus (and its synonymous forms such as Oran, Bermejo, and Lechiguanas viruses) and especially the Rio Mamoré (RM) virus of *Oligoryzomys microtis*, with which LN virus may be synonymous (Fig. 1; Hjelle et al. 1996; López et al. 1996; Bharadwaj et al. 1998; Levis et al. 1998). The host for LN virus, however, is the phyllotine species *Calomys laucha*, a distant relative of the oryzomine rodents within the subfamily Sigmodontinae. In the Old World, it has been noted, Topografov virus of Siberian lemmings and Khabarovsk virus of *Microtus fortis* are much more closely related than are their respective rodent hosts (Vapalahti et al. 1999). Thus, hantavirus evolutionary behavior generally reflects deep evolutionary relationships among rodent hosts as equally deep viral phylogenetic relationships when given the luxury of prolonged association with their rodent hosts. When rodent evolution occurs rapidly, hantaviruses presumably also must speciate rapidly, and opportunistic host switching is more likely to be observed.

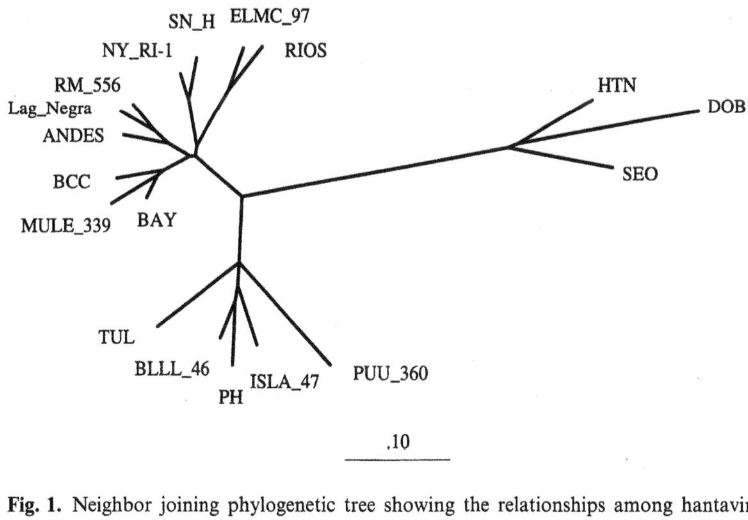

Fig. 1. Neighbor joining phylogenetic tree showing the relationships among hantaviruses for which a complete nucleocapsid gene sequence is available. The tree shown is based upon the 428–433 amino acid nucleocapsid protein sequence. A *scale bar* indicates a 10% amino acid distance. The cluster of viruses from the *upper left* is derived from rodents of the subfamily Sigmodontinae, family Muridae. Those from the *bottom* cluster are vole-borne viruses (subfamily Arvicolinae, family Muridae). Those from the *upper right* are from rodents of the subfamily Murinae, family Muridae

Purifying selection during hantavirus evolution appears to occur primarily at the amino acid level (Hjelle et al. 1995; López et al. 1997). There is an extremely high ratio of synonymous to nonsynonymous substitution when hantaviruses of a given species are compared across a large geographic space. Variation of 18–20% at the nucleotide level may translate to changes of only 2–3% at the amino acid level (Hjelle et al. 1995). An exception is Puumala virus, which appears to be subject to a somewhat less-intense purifying selection in nature. It is only when hantaviruses from different reservoir hosts are compared, such as when comparing the sympatric SN and El Moro Canyon viruses of the American southwest, that amino acid sequences begin to diverge substantially.

1.1 Lessons from "Spillover" Infections

The "true" reservoir host for a particular hantavirus can be determined by establishing what species of rodent is most frequently infected with a particular species of virus within a given geographic region. Serologic testing can be used to establish the prevalence for hantaviruses within each sympatric rodent species; there is little or no evidence for an antibody negative carrier state within rodents such as may be observed with arenaviruses. Usually one species of rodent is heavily over-represented among seropositive specimens in field surveys (Childs et al. 1994; Niklasson et al. 1995). Historically, however, there has been a tendency among investigators to attribute all positive antibody responses to whatever virus antigen is in use in their diagnostic assays. Actually, anti-hantavirus antibodies frequently cross-react with multiple hantavirus serotypes. As an example, the existence of El Moro Canyon virus, which occurs in the Four Corners region where SN virus was first recognized, was obscured when seropositive but PCR-"negative" specimens of *R. megalotis* were considered to represent spillover infection by SN virus.

Spillover infections are infections of rodent hosts other than the species identified as the predominant carrier for a particular virus. In the southwestern United States, where SN virus occurs in complex rodent communities, *P. maniculatus* is the predominant carrier species in most areas. Often, *P. maniculatus* is more abundant than other rodent species and the seroprevalence of SN virus antibodies is highest in *P. maniculatus*. In some areas, other congenerics such as *P. boylii* have higher prevalence, even if their population density is lower than that of *P. maniculatus* (Mills et al. 1997). *P. boylii* is now known to carry a distinct hantavirus that resembles El Moro Canyon virus more so than SN, perhaps reflecting another example of recent host switching (S. Nichol, personal communication; B. Hjelle, unpublished data). Other rodents in the community are carriers of SN virus. The more closely related these rodents are to the most prominent carrier, *P. maniculatus*, the more likely they are to have SN virus antibodies, and the more likely they are to have viral RNA detectable in their tissues using the reverse transcription-polymerase chain reaction (RT-PCR; Childs et al. 1994). However, the vast majority of antibody-positive animals do have viral RNA in their tissues, even in spillover infections (B. Hjelle, unpublished data). The inability to detect viral RNA in some spillover infections may be related to the sensitivity of the RT-PCR assay.

Spillover infections are much more common in nature than is generally appreciated. Multiple hantaviruses circulate among wild rodents in many parts of the world. We have encountered, for example, the presence of viruses adapted for harvest mice in deer mice and vice-versa, and vole-associated viruses in deer mice, white-footed mice, and cotton rats, all sigmodontine rodents (Song et al. 1995; Hjelle et al. 1995; Rawlings et al. 1996; Torrez-Martinez et al. 1998; B. Hjelle et al., unpublished data). In the great majority of cases, viral RNA can be detected in the tissues of the spillover hosts, indicating that most likely the virus is able to establish a chronic infection. Co-infections in which more than one species of hantavirus infects a single rodent specimen have not been described.

Despite the cross-infection of host species with different hantaviruses, there is no evidence that different species of hantavirus may exchange (reassort) genomic segments in nature. Intraspecific reassortment has been observed convincingly for SN virus and less-convincing examples have been noted for other hantaviruses (Li et al. 1995). Even in regions where several hantaviruses are sympatric and have cross-infected the same rodent species, natural reassortment between species has been lacking, and attempts to produce experimental reassortants in the laboratory have been generally unsuccessful (Rawlings et al. 1996; Rodriguez et al. 1998). Whether using tissue culture systems or observing natural animal infections, it appears that the segments of a virus of a particular serotype produce a virus of optimal fitness only when they occur together. To our knowledge, deliberate attempts to induce interspecific segment reassortment by infecting a single rodent simultaneously with more than one hantavirus have not been conducted.

1.2 Modeling Determinants of Host Tropism

Each hantavirus appears to be closely adapted for survival with its predominant reservoir host. The selective influences that enforce this adaptation are not understood. Presumably viral replication, shedding of infectious virus in one or more bodily fluids, immune escape, and perhaps periodic reactivation of shedding are important in maintenance of the virus in rodent populations. Because spillover infections of alternate rodent hosts may occur, associated with sufficient replication for routine detection by RT-PCR, the specificity of virus for host is not absolute. Clearly there is a substantial difference between the barriers that inhibit the successful propagation of a virus species in an alternative host, and the long-term, independent success of the virus in such a host. The lack of frequent host switching by hantaviruses would suggest that either the number of changes required to adapt to a new species are only rarely generated in the course of hantavirus replication within a host, that such variants are only rarely passed on to new rodent host species, or both. If a new carrier rodent species already were a hantavirus reservoir, a variant that has been newly adapted to the host would face considerable ecological competition from the pre-existing virus.

In addition to naturally occurring transient infections of humans (Hjelle et al. 1994), where viral RNA declines to undetectable levels shortly after recovery, some hantaviruses will at least transiently infect hamsters, gerbils or laboratory mice (*M. musculus*). These models have been used for production of vaccine preparations as well as for studies of the protective efficacy of vaccines (Yanagihara et al. 1985; Schmaljohn et al. 1990; Xu et al. 1992; Chu et al. 1995).

1.3 Pathologic Consequences of Hantavirus Infection in Rodent Hosts

Hantaviruses have long been thought to cause no detrimental clinical effects on infected reservoir host rodents (Childs et al. 1989). That hantaviruses do not cause hemorrhagic fever with renal syndrome (HFRS) or hantavirus cardiopulmonary syndrome (HCPS) in rodents has been a disappointment, since it means that we have no animal model for the human hantavirus-associated diseases. Some hantaviruses cause a fatal infection of the central nervous system when inoculated intracranially into inbred *M. musculus* (Tsai et al. 1982). The evidence that hantaviruses do not harm their reservoir hosts comes from studies that have shown that seropositive animals appear to have approximately the same lifespan as seronegative animals. Laboratory infection models have not shown clinical illness upon experimental infection of any rodent host and pulmonary physiologic testing of wild-caught deer mice infected with SN virus have not shown any evidence of respiratory insufficiency (Lee et al. 1981; O'Connor et al. 1997; Hutchinson et al. 1998).

Recently, two groups have reported that reservoir rodents seropositive for SN virus or the related New York virus have an increased prevalence of pulmonary edema and periportal hepatitis, and raise the possibility that at least histopathological stigmata compatible with mild HCPS can occur in wild, infected rodents (Lyubski et al. 1996; Netski et al. 1999). These studies have been difficult to interpret due to difficulties with obtaining tissues of uninfected control rodents that were of the same genetic background and that were treated similarly to the tissues of infected rodents.

1.4 Animal Models for Hantavirus Infection: Why Are They Difficult?

It is remarkable that animal models for some hantavirus infections are rather difficult to develop, given the high natural prevalence of infection in many field studies. Although viral antigens and RNA are present in the tissues of most seropositive rodents, the viruses are often very difficult to establish in tissue culture or in animal infections by direct inoculations. Many explanations have been offered. In some cases, it appears that the hantavirus genome must undergo specific mutations during the passage to tissue culture (Lundkvist et al. 1997). A similar rate-limiting genetic adaptation may be required for the re-adaptation of the virus from Vero E6 cells in vitro for passage into animals. For this reason, it may be advisable to avoid a tissue culture adaptation step for experimental passage of some hantaviruses. Other hantaviruses appear unconstrained in passage between tissue culture and animals (HUTCHINSON et al. 1998).

Other factors may be limiting the ability to experimentally infect rodent hosts with hantaviruses. For example, it is possible that a reservoir host from one geographic region may be less permissive for infection with a hantavirus obtained from another region. The presence of neutralizing antibodies, defective interfering viruses, co-infections, host genetic lesions that impede viral entry or replication, or variations in innate immunity among different individual host rodents may be important factors in some instances. Since it is not known how hantaviruses are generally passed among members of the reservoir host species, the route of administration (intramuscular, intranasal, subcutaneous, or intratracheal) may be an important factor as well.

The requirement for biosafety level 3 or 4 containment for propagation of the viruses associated with hantavirus cardiopulmonary syndrome (HCPS) severely restricts the number of laboratories that can develop animal models for infection with these agents (CENTERS FOR DISEASE CONTROL AND PREVENTION 1994). Fortunately, it has been possible recently to develop outdoor laboratories that can carry out infection with these highly pathogenic agents, using outdoor biosafety standards (MILLS et al. 1995; BOTTEN et al. 2000a,b). More widespread use of such innovative solutions may allow more laboratories to participate in work with certain level-4 agents (Fig. 2).

Fig. 2. Outdoor biocontainment facility near Socorro, New Mexico. A worker wearing a powered air-purifying respirator (PAPR) is shown working in a novel outdoor biocontainment laboratory. Infected and uninfected deer mice are maintained in this outdoor facility inside individual cages. In each cage, the mice are provided an artificial burrow that prevents escape and allows them to retreat from summer heat, precipitation, and predation (BOTTEN et al. 2000a). The use of such a facility allows one to safely maintain hantavirus-infected mice without the need for biosafety level-4 containment. This facility was recently used to establish the first deer mouse model for infection by Sin Nombre hantavirus

2 Models of Enzootic and Epizootic Transmission of Hantaviruses in Rodent Communities: Effects of Ecological Factors on Rodent Populations

The order Rodentia contains the most species of mammalian orders and has members that are keystone species in most ecosystems of the world. Much of the ecological complexity of this diverse order can be found within one family, the Muridae. Murid rodents occur worldwide and occupy essentially all major habitats. The family is highly variable from a population standpoint with some members maintaining relatively stable population densities while others vary widely both in terms of population density and local distribution. This family contains species that serve as reservoirs for all but one known species of hantavirus and also all but one species of arenavirus. It is these murid species in particular that are subject to large population changes, either as a result of human land-use practices, climate change, or normal population cycles, that pose the most serious threats to human health.

It is often difficult to separate climate change from normal climatic fluctuation, but both are known to affect rodent populations. Rapid fluctuations in murid rodent population densities are particularly pronounced in semiarid and arid environments but may occur in any habitat given the correct set of environmental variables. Increased precipitation associated with large-scale phenomena such as the El Niño Southern Oscillation (ENSO) has been suggested as a possible cause of increased rodent population densities in both South (Meserve et al. 1993: Jaksic et al. 1997) and North America (Yates et al. 1998) that were, in turn, associated with an increased caseload of human cardiopulmonary syndrome (Hjelle and Glass 2000). There are also decreases in rodent population densities associated with the drought conditions affiliated with the La Niña side of the ENSO. Different murid species, however, respond differently to the same environmental change even at the same location (Meserve et al. 1995) or differently at different altitudes or latitudes (Abbott et al. 1999; Calisher et al. 1999; Kuenzi et al. 1999; Mills et al. 1999; Parmenter et al. 1999). Another group of murid rodents, the Arvicolinae, undergo fairly regular population increases and decreases on 3- to 4-year cycles. The causes of these cycles are poorly understood but at least in some cases food quality and quantity may be involved (Tast and Kalela 1971; Niklasson et al. 1995), and populations may be intrinsically programmed to reduce breeding effort when population densities are high (Ostfeld et al. 1993).

Human activities have considerable impact on the abundance and distribution of plant and animal species and are especially important to population dynamics of murid rodent species. Anthropogenic ecosystem disturbances, resulting from mining, grazing, deforestation, agriculture, construction of human communities and roads, and other activities of man have altered the composition of biotic and abiotic resources worldwide. These changes often favor murid rodents over other species resulting in a reduction of species in the community but higher densities of those that remain. This can be particularly dangerous where hantavirus is concerned because increased population densities often lead to increased contact between humans and infected mice. Increases in rodent populations which are associated with increased human disease have been associated with fall harvests in Argentina (Mills et al. 1992) and Korea (LeDuc 1989), for example, and much larger numbers of hantavirus-infected mice were found in areas in Brazil that were converted to agriculture than in adjacent undisturbed habitats (E. Burt, personal communication).

In addition to increasing the food supply, these activities create an ideal habitat for many New World species of the family Muridae, since much of the early evolutionary history of this group was in the Miocene grasslands of North America. This seems to be particularly true when primary forests are clear-cut for housing. Rodents require a number of habitat resources including food, shelter, and reproductive sites. This includes vertical architecture (grass vs shrubs vs trees) and horizontal architecture (patch size and spatial distribution of vegetation types). Because rodents vary tremendously in their vagility and ability to utilize specific resources in the environment, generalists such as many hantavirus reservoirs will be favored in less species-rich and ecologically less complex communities such as grasslands. In addition, many of the species that carry hantaviruses are easily able to utilize manmade buildings, especially during times of high population density, when competition for resources is high and many animals are forced to disperse. Populations may increase in density even further during the favorable climatic periods mentioned above.

Remarkably little is known about the mechanisms for maintenance of hantavirus infections among reservoir rodent populations. It is clear, however, that infection levels change rapidly as population density in rodent populations fluctuates. Mills et al. (1997) found populations with seropositive murid rodents in virtually all biomes in the southwestern part of North America. Seroprevalences were found by these authors to vary among populations and species. Of the three most common species of sigmodontine rodents examined, deer mice exhibited the lowest average (11%) fraction of seropositive specimens. Local population seroprevalences vary greatly across geographic space and with time. For example, Mills et al. (1997) found infection rates in local reservoir populations to range from 0 to 47%, but found no correlation between population density and seroprevalences. Abbott et al. (1999), however, found the number of antibody positive brush mice to be higher at study sites in Arizona during periods of high density. Other authors have found a decrease in seropositive *Peromyscus* when population densities become extremely low (Abbott et al. 1999; Parmenter et al. 1999).

The seroprevalence of hantaviruses in wild populations of rodents may well be determined in part by random processes. Given the average seroprevalence of 11% reported by Mills et al. (1997), for example, it would not be surprising to see local extinction of hantavirus in rodents populations due to random chance as populations decrease, especially if the decline in population density occurs rapidly. Assuming mice are removed from the population at random, those that remain in small isolated populations may or may not have been those that were infected. If so, then widespread hantavirus infection in virus-free rodent populations would require reinfection by individuals that have dispersed from populations that had maintained the virus since the population crash. Whether infection is always present in the same refugia populations or is maintained in different refugia each time there is a major change in population density is an interesting question yet to be investigated, as is the mechanism by which hantaviruses are transmitted in the wild.

Attempts to determine the mechanisms for transmission of virus among reservoir host species have involved both provocative laboratory experiments involving animal models or field studies involving wild-caught rodent specimens (LEE et al. 1981; GAVRILOVSKAYA et al. 1990; MILLS et al. 1997; BENNETT et al. 1999). Laboratory studies of intracage transmission are highly artificial, and their relevance to infections in natural settings is questionable. Such studies have demonstrated that in indoor settings, rodents may transmit virus to cagemates and across airspace between cages. In experimental infection models, virus is shed in the urine, feces, and saliva, especially in the acute phase of illness (LEE et al. 1981; YANAGIHARA et al. 1985).

Field studies have demonstrated that hantavirus seroprevalence is highly focal, with some rodent communities exhibiting as little as 0–1% and others as much 50% prevalence (MILLS et al. 1997; BENNETT et al. 1999; B. Hjelle, unpublished data). Seroprevalence at a particular site can vary greatly over time. For most hantaviruses, larger male specimens are much more commonly infected than smaller males or females (HJELLE et al. 1995; MILLS et al. 1997; BENNETT et al. 1999). Norwegian rats are an exception; Seoul virus antibody prevalence is similar between males and females (CHILDS et al. 1987). Seoul virus seroprevalence among wild rats increases with body mass, and animals with deep wounds are especially commonly infected. This observation has led to the belief that hantaviruses are transmitted largely through biting, fighting, or other aggressive behaviors, perhaps by direct intramuscular or subcutaneous inoculation of virus-contaminated saliva (GLASS et al. 1988). Direct confirmatory evidence of this hypothesis has been lacking. It is plausible that animals that display aggressive tendencies and acquire wounds are also more likely to enter burrows where contaminated excreta may be found, or that such animals are more likely to be exposed to virus through sexual exposure to infected mates. Wounding may also be an indicator of personal traits that result in long-term survival in the wild, with associated greater lifetime exposure to virus by a number of routes.

It is easy to imagine how hantaviruses are maintained in rodent populations during times of high resource availability, high population density, and active breeding. However, during periods of low food availability and declining population densities, hantaviruses remain able to avoid extinction. In some settings, a few large adults, which may have been infected for several months, remain the only source of virus (B. Hjelle et al., unpublished data). When rodent populations recover, virus must either re-enter the population from distant refugia, or the virus must be reactivated from a relatively quiescent state, as is seen in chronic infection. It is tempting to speculate that hantaviruses have evolved a mechanism that exploits hormonal or environmental signals to reactivate as populations recover from a cycle of poor productivity. In this light, it is interesting that hantaviruses appear to use brown fat as an important depot in infected hosts (GAVRILOVSKAYA et al. 1983).

This tissue exhibits abundant expression of antigen in deer mice that have been experimentally infected by SN virus (Botten et al. 2000b).

3 Need for Interdisciplinary Study

Investigators engaged in field studies of hantaviruses are handicapped by their minimal ability to perturb the virus-host system to dissect the effects of individual changes. Nor is it possible to control the multitude of ecological variables present in the field. Laboratory studies such as infection models are limited by their remote relationship to natural conditions, and by the small scale of the laboratory environment. Real understanding of the mechanisms of hantavirus maintenance will require interdisciplinary studies that bridge the gap between provocative laboratory studies and field investigations. Systems that combine the use of field and laboratory models, for example, in carefully manipulated and monitored field enclosures in otherwise undisturbed habitat, may allow more detailed inferences about the mechanisms that maintain hantaviruses in nature and the investigation of the natural complexity of infectious diseases in nature. Experiments of such complexity will require the development not only of new field laboratories, but also the development of multidisciplinary teams that can speak the language of each individual discipline. Such efforts hold great promise for breakthroughs in the understanding of the ecology of infectious diseases.

4 Concluding Remarks

Studies of the ecology of hantavirus infections in the wild are coming of age with the recent application of molecular methods, more in-depth study of virus-host relationships, and innovative new tools for ecological modeling and biological containment and the advent of truly multidisciplinary long-term studies. These innovations in the study of hantavirus ecology should spur similar innovation in studies of other zoonotic diseases, ranging from arenavirus diseases to Lyme disease and Ehrlichiosis. The effort is at a nascent stage, however, and the power of such technology will not be fully realized for several years.

While the innovative strategies being developed to study the hantaviruses has been exciting, it is of some concern that the wild reservoir species that are the natural carriers for zoonotic diseases have not been developed sufficiently as animal models. Long-term scientific interest has traditionally focused upon a very limited number of animal models that are reservoirs for very few diseases, such as laboratory mice, guinea pigs, *Rattus*, and hamsters. Researchers and resources have tended to gravitate toward those animal models for which the most reagents are available, and interest has been slight in the development of models employing natural disease reservoirs. The relevant host-pathogen relationships involve species such as wild birds, bats, and rodents. While development of such models may be slower and more difficult, the long-term interest of controlling zoonotic diseases will require more extensive use and characterization of such host species.

The beginning of the merger of the fields of public health, epidemiology, and virology, with those of ecology, systematics, and evolution promises a major conceptual advance in our efforts to understand and model zoonotic diseases. It has become increasingly clear that many of the factors driving the emergence of infectious diseases today are far more subtle and occur on a much larger scale than previously believed (Yates 1994). A rapidly increasing human and domestic animal population, coupled with global climate change and an increasingly mobile world population, require the study of emerging infections on a planetary scale. The solutions of the future will require large-scale, multidisciplinary, long-term, and international efforts.

References

Abbott KD, Ksiazek TG, Mills JN (1999) Long-term hantavirus persistence in rodent populations in central Arizona. Emerg Infect Dis 5:102–112

Baskin JA (1989) The initial origin and diversification of the Neotropical Sigmodontinae (Rodentia: Muridae) – a perspective from the North American fossil record. In: Abstracts of papers and posters, Fifth International Theriological Congress, Rome, pp 263–264

Bennett SG, Webb JP Jr, Madon MB, Childs JE, Ksiazek TG, Torrez-Martinez N, Hjelle B (1999) Hantavirus (*Bunyaviridae*) infections in rodents from Orange and San Diego Counties, California. Am J Trop Med Hyg 60:75–84

Bharadwaj M, Botten J, Torrez-Martinez N, Hjelle B (1998) Rio Mamoré virus: genetic characteristics of a newly recognized hantavirus of the pygmy rice rat *Oligoryzomys microtis* from Bolivia. Am J Trop Med Hyg 57:368–374

Botten J, Nofchissey R, Ahern H, Rodriguez-Moran P, Wortman IA, Goade D, Yates T, Hjelle B (2000a) Outdoor facility for quarantine of wild rodents infected with hantavirus. J Mammal 81:250–259

Botten J, Mirowsky K, Kusewitt D, Bharadwaj M, Yee J, Ricci R, Feddersen RM, Hjelle B (2000b) Experimental infection model for Sin Nombre hantavirus in the deer mouse (*Peromyscus maniculatus*). Proc Natl Acad Sci USA 97:10578–10583

Calisher CH, Sweeney W, Mills JN, Beaty BJ (1999) Natural history if Sin Nombre Virus in western Colorado. Emerging Inf Dis 5(1):126–134

Carey D, Reuben R, Panicker K, Shope R, Myers R (1971) Thottapalayam virus: a presumptive arbovirus isolated from a shrew in India. Indian J Med Res 59:1758–1760

Centers for Disease Control and Prevention (1994) Laboratory management of agents associated with hantavirus pulmonary syndrome: interim biosafety guidelines. Morb Mortal Weekly Rep 43: RR-7, 1–7

Childs JE, Korch GW, Glass GE, LeDuc JW, Shah KV (1987) Epizootiology of hantavirus infections in Baltimore: isolation of a virus from Norway rats, and characteristics of infected rat populations. Am J Epidemiol 126:55–68

Childs JE, Glass GE, Korch GW, LeDuc JW (1989) Effects of hantaviral infection on survival, growth and fertility in wild rat (*Rattus norvegicus*) populations of Baltimore, Maryland. J Wildlife Dis 25:469–476

Childs JE, Ksiazek TG, Spiropoulou CF, Krebs JW, Morzunov S, Maupin GO, Gage KL, Rollin P, Sarisky J, Enscore R, Peters CJ, Nichol ST (1994) Serologic and genetic identification of *Peromyscus maniculatus* as the primary rodent reservoir for a new hantavirus in the southwestern United States. J Infect Dis 169:1271–1280

Chu YK, Jennings GB, Schmaljohn CS (1995) A vaccinia virus-vectored Hantaan virus vaccine protects hamsters from challenge with Hantaan and Seoul viruses but not Puumala virus. J Virol 69:6417–6423

Gavrilovskaya IN, Apekina NS, Myasnikov YuA, Bernshtein AD, Ryltseva EV, Gorbachkova EA, Chumakov MP (1983) Features of circulation of hemorrhagic fever with renal syndrome (HFRS) virus among small mammals in the European USSR. Arch Virol 75:313–316

Gavrilovskaya IN, Apekina NS, Bernshtein AD, Demina VT, Okulova NM, Myasnikov YA, Chumakov MP (1990) Pathogenesis of hemorrhagic fever with renal syndrome virus infection and mode of horizontal transmission of hantavirus in bank voles. Arch Virol 91:57–62

Glass GE, Childs JE, Korch GW, LeDuc JW (1988) Association of intraspecific wounding with hantaviral infection in wild rats (Rattus norvegicus). Epidemiol Infect 101:459–472

Hjelle B, Spiropoulou CF, Torrez-Martinez N, Morzunov S, Peters CJ, Nichol ST (1994) Detection of Muerto Canyon virus RNA in peripheral blood mononuclear cells from patients with hantavirus pulmonary syndrome. J Infect Dis 170:1013–1017

Hjelle B, Anderson B, Torrez-Martinez N, Song W, Gannon WF, Yates TL (1995) Prevalence and geographic genetic variation of hantaviruses of New World harvest mice (*Reithrodontomys*): identification of a divergent genotype from a Costa Rican *Reithrodontomys mexicanus*. Virology 207:452–459

Hjelle B, Torrez-Martinez N, Koster FT (1996) Hantavirus pulmonary syndrome-related virus from Bolivia. Lancet 347:57

Hjelle B, Glass GE (2000) Outbreak of hantavirus infection in the Four Corners region of the US in the wake of the 1997–98 El Niño-Southern Oscillation. J Infect Dis 181:1569–1573

Hutchinson KL, Rollin PE, Peters CJ (1998) Pathogenesis of a North American hantavirus, Black Creek Canal virus, in experimentally infected *Sigmodon hispidus*. Am J Trop Med Hyg 59:58–65

Jaksic FM, Silva SI, Meserve PL, Gutierrez JR (1997) A long-term study of vertebrate predator responses to an El-Nino (ENSO) disturbance in western South America. Oikos, 78(2):341–354

Johnson AM, Bowen MD, Ksiazek TG, Williams RJ, Bryan RT, Mills JN, Peters CJ, Nichol ST (1997) Laguna Negra virus associated with HPS in western Paraguay and Bolivia. Virology 238:115–127

Kuenzi AJ, Morrison ML, Swann DE, Hardy PC, Downard GT (1999) A longitudinal study of Sin Nombre Virus prevalence in rodents, southeastern Arizona. Emerging Inf Dis 5(1):113–117

LeDuc JW (1989) Epidemiology of Hemorrhagic Fever viruses. Rev Infect Dis 11:S730

Lee HW, Lee PW, Baek LJ, Song CK, Seong IW (1981) Intraspecific transmission of Hantaan virus, etiologic agent of Korean hemorrhagic fever, in the rodent *Apodemus agrarius*. Am J Trop Med Hyg 30:1106–1112

Levis S, Morzunov SP, Rowe JE, Enria D, Pini N, Calderon G, Sabattini M, St. Jeor SC (1998) Genetic diversity and epidemiology of hantaviruses in Argentina. J Infect Dis 177:529–538

Li D, Schmaljohn AL, Anderson K, Schmaljohn CS (1995) Complete nucleotide sequences of the M and S segments of two hantavirus isolates from California: evidence for reassortment in nature among viruses related to hantavirus pulmonary syndrome. Virology 206:973–983

López N, Padula P, Rossi C, Lazaro ME, Franze-Fernandez MT (1996) Genetic identification of a new hantavirus causing severe pulmonary syndrome in Argentina. Virology 220:223–226

López N, Padula P, Rossi C, Miguel S, Edelstein A, Ramirez E, Franze-Fernández MT (1997) Genetic characterization and phylogeny of Andes virus and variants from Argentina and Chile. Virus Res 50:77–84

Lundkvist A, Cheng Y, Sjolander KB, Niklasson B, Vaheri A, Plyusnin A (1997) Cell culture adaptation of Puumala hantavirus changes the infectivity for its natural reservoir, *Clethrionomys glareolus*, and leads to accumulation of mutants with altered genomic RNA S segment. J Virol 71:9515–9523

Lyubsky S, Gavrilovskaya I, Luft B, Mackow E (1996) Histopathology of *Peromyscus leucopus* naturally infected with pathogenic NY-1 hantaviruses: pathologic markers of HPS viral infection in mice. Lab Invest 74:627–633

Mertz GJ, Hjelle BL, Bryan RT (1997) Hantavirus infection. Adv Intern Med 42:369–421

Meserve PL, Yunger JA, Gutierrez JR, Contreras LC, Milstead WB, Lang BK, Cramer KL, Herrera S, Lagos VO, Silva SL, Tabilo EL, Torrealba MA, Jaksic FM (1995) Hetrogeneous responses of small mammals to an El-Nino Southern-Oscillation event in northcentral semiarid Chile and the importance of ecological scale. J Mamm 76(2):580–595

Meserve PL, Gutierrez JR, Contreras LC, Jaksic FM (1993) Role of Biotic Interactions in a semiarid scrub community north-central Chile: a long-term ecological experiment. Revista Chilena de Historia Natural 66(3):225–241

Mills JN, Ellis BA, McKee KT Jr, Calderon GE, Maiztegui JI, Nelson GO, Ksiazek TG, Peters CJ, Childs JE (1992) A longitudinal study of Junin virus activity in the rodent reservoir of Argentine hemorrhagic fever. Am J Trop Med Hyg 47:749–763

Mills JN, Yates TL, Childs JE, Parmenter RR, Ksiazek TG, Rollin PE, Peters CJ (1995) Guidelines for working with rodents potentially infected with hantavirus. J Mammal 76:716–722

Mills JN, Ksiazek TG, Ellis BA, Rollin PE, Nichol ST, Yates TL, Gannon WL, Levy CE, Engelthaler DM, Davis T, Tanda DT, Frampton JW, Nichols CR, Peters CJ, Childs JE (1997) Patterns of association with host and habitat: antibody reactive with Sin Nombre virus in small mammals in the major biotic communities of the southwestern US. Am J Trop Med Hyg 56:273–284

Mills JN, Ksiazek TG, Peters CJ, Childs JE (1999) Long-term studies of hantavirus reservoir populations in the Southwestern United States: a synthesis. Emerging Inf Dis 5(1):135–142

Netski D, Thran BH, St Jeor SC (1999) Sin Nombre virus pathogenesis in *Peromyscus maniculatus*. J Virol 73:585–591

Nichol ST (in press) Genetic analysis of hantaviruses and their host relationships. In: Saluzzo JF, Dodet B (eds) Emergence and control of rodent-borne viral diseases. Elsevier, Paris

Niklasson B, Hornfeldt B, Lundkvist A, Bjorsten S, LeDuc J (1995) Temporal dynamics of Puumala virus antibody prevalence in voles and of nephropathia epidemica incidence in humans. Am J Trop Med Hyg 53:134–140

O'Connor CS, Hayes JP, St. Jeor SC (1997) Sin Nombre virus does not impair respiratory function of wild deer mice. J Mammal 78:661–668

Ostfeld RS, Canham CD, Pugh SR (1993) Intrinsic density-dependent regulation of vole populations. Nature 366:259–261

Parmenter CA, Yates TL, Parmenter RR, Dunnum JL (1999) Statistical sensitivity for detection of spatial and temporal patterns in rodent population densities. Emerging Inf Dis 5(1):118–125

Rawlings JA, Torrez-Martinez N, Neill SU, Moore GM, Hicks BN, Pichuantes S, Nguyen A, Bharadwaj M, Hjelle B (1996) Cocirculation of multiple hantaviruses in Texas, with characterization of the S genome of a previously undescribed virus of cotton rats (*Sigmodon hispidus*). Am J Trop Med Hyg 55:672–679

Reig OA (1978) Roedores cricetidos del plioceno superior de la provincia Buenos Aires (Argentina). Publ Mus Minicipal Cienc Nat Lorenzo Scaglia Mar del Plata 2:164–190

Rodriguez LL, Owens JH, Peters CJ, Nichol ST (1998) Genetic reassortment among viruses causing hantavirus pulmonary syndrome. Virology 242:99–106

Schmaljohn CS, Chu YK, Schmaljohn AL, Dalrymple JM (1990) Antigenic subunits of Hantaan virus expressed by baculovirus and vaccinia virus recombinants. J Virol 64:3162–3170

Schmaljohn C, Hjelle B (1997) Hantaviruses: a global disease problem. Emerg Infect Dis 3:95–104

Song W, Torrez-Martinez N, Irwin W, Harrison J, Davis R, Ascher M, Jay M, Hjelle B (1995) Isla Vista virus: a genetically novel hantavirus of the California vole *Microtus californicus*. J Gen Virol 76: 3195–3199

Tast J, Kalela O (1971) Comparisons between rodent cycles and plant production in Finnish Lapland. Ann Acad Sci Fenn A, IV Biologica:186

Torrez-Martinez N, Bharadwaj M, Goade D, Delury J, Moran P, Hicks B, Nix B, Davis J, Hjelle B (1998) Bayou virus-associated hantavirus pulmonary syndrome in eastern Texas: identification of *Oryzomys palustris* as the reservoir host. Emerg Infect Dis 4:105–111

Tsai TF, Bauer S, McCormick JB, Kurata T (1982) Intracerebral inoculation of suckling mice with Hantaan virus. Lancet 2:503–504

Vapalahti O, Lundkvist A, Fedorov V, Conroy CJ, Hirvonen S, Plyusnina A, Nemirov K, Fredga K, Cook JA, Niemimaa J, Kaikusalo A, Henttonen H, Vaheri A, Plyusnin A (1999) Isolation and characterization of a hantavirus from *Lemmus sibiricus*: evidence for host switch during hantavirus evolution. J Virol 73:5586–5592

Xu X, Ruo SL, McCormick JB, Fisher-Hoch SP (1992) Immunity to Hantavirus challenge in *Meriones unguiculatus* induced by vaccinia-vectored viral proteins. Am J Trop Med Hyg 47:397–404

Yanagihara R, Goldgaber D, Gajdusek DC (1985) Propagation of nephropathia epidemica virus in Mongolian gerbils. J Virol 53:973–975

Yates TL (1994) Assessing and monitoring biological diversity at the genetic level. In: Watanabe MM, Yasuno M (eds) Proceedings of the international workshop on biodiversity. National Institute for Environmental Studies, Tsukuba, Japan, pp 41–49

Yates TL, Ksiazek T, Parmenter R, Rollin P, Nichol S, Dunnum J, Baker R, Parmenter C, Peters C (1998) Hantavirus outbreaks and rodent ecology: the role of El Nino. In: Abstracts of the Fourth International Conference on HFRS and Hantaviruses, p 61

Cellular Receptors and Hantavirus Pathogenesis

E.R. Mackow[1] and I.N. Gavrilovskaya[2]

[1] Department of Medicine, Stony Brook University, HSC T17, Room 60, Stony Brook, NY 11794, USA
[2] Northport VA Medical Center, Northport NY 11768, USA

1 Introduction

Hantaviruses cause two frequently fatal human diseases, hemorrhagic fever with renal syndrome (HFRS) and hantavirus pulmonary syndrome (HPS) (NOLTE et al. 1995; ZAKI et al. 1995; GONZALEZ-SCARANO et al. 1996; SCHMALJOHN 1996). HFRS-associated hantaviruses cause vascular hemorrhage and kidney dysfunction while HPS-associated hantaviruses cause acute pulmonary edema (LEE et al. 1999; ZAKI et al. 1999). Both diseases are associated with acute thrombocytopenia and changes in vascular permeability, and either disease may have pulmonary or renal components (NOLTE et al. 1995; ZAKI et al. 1995; GONZALEZ-SCARANO and NATHANSON 1996; SCHMALJOHN 1996; LEE et al. 1999; ZAKI and NOLTE 1999).

The means by which specific hantaviruses cause pulmonary or renal diseases and increase vascular permeability is just beginning to be investigated. Hantaviruses cause disease in humans but not in their animal hosts, and both pathogenic and non-pathogenic hantaviruses have the same tissue tropism, replicating predominantly in endothelial cells and macrophages (YANAGIHARA and SILVERMAN 1990; PENSIERO et al. 1992; HUGHES et al. 1993; NOLTE et al. 1995; ZAKI et al. 1995; GONZALEZ-SCARANO and NATHANSON 1996). Infected endothelial cells are not lysed by hantaviruses, and although immune cells are recruited to the infected endothelium (ZAKI et al. 1995), it is unclear to what extent these contribute to vascular disease (YANAGIHARA and SILVERMAN 1990; NOLTE et al. 1995). Recent findings indicate that pathogenic and non-pathogenic hantaviruses interact with discrete integrin receptors present on platelets, immune and endothelial cells (GAVRILOVSKAYA et al. 1998; GAVRILOVSKAYA et al. 1999; MACKOW et al. 1999). These findings link hantavirus pathogenesis to the use of β3 integrins, which are central to the regulation of platelet function and vascular permeability (GAVRILOVSKAYA et al. 1998; GAVRILOVSKAYA et al. 1999; MACKOW et al. 1999).

2 Hemorrhagic Fever with Renal Syndrome

HFRS was first described over 50 years ago, and hantaviruses were first isolated in 1978 (LEE et al. 1978; SCHMALJOHN et al. 1997). Hantaviruses derive their name from the Hantaan river in Korea where Hantaan virus, the prototype HFRS-causing hantavirus, was found to be the etiologic agent of Korean hemorrhagic fever (LEE et al. 1978). Hantaan (HTN), Seoul (SEO), Puumala (PUU) and Dobrava (DOB) viruses are prominent causes of human HFRS and induce various degrees of disease with 0.1–5% mortality rates (LEE 1982; COSGRIFF et al. 1991; AVSIC-ZUPANC et al. 1992). HFRS disease is characterized by an influenza-like illness with microvascular hemorrhage, platelet loss, hypotension, shock, and in some cases renal failure (LEE 1982; COSGRIFF and LEWIS 1991; AVSIC-ZUPANC et al. 1992). Activation of complement cascades in HFRS patients supports a role for accumulated immune complexes in renal disease (Gavrilovskaya et al. 1987). However, the mechanism by which HFRS viruses cause vascular hemorrhage, thrombocytopenia, and kidney disease are not well understood.

Hantaan virus is estimated to cause over 100,000 cases of HFRS annually in Asia. Seoul virus is present worldwide, although reported cases occur predominantly within Asia. Puumala is a substantial cause of HFRS morbidity within Europe, and Dobrava virus is a newly identified cause of severe HFRS within the Balkan region of Europe. Although PUU infections are characterized by a low mortality rate, this and other hantaviruses are severely debilitating and require prolonged recovery periods lasting from months to more than a year. A more complete description of hantavirus clinical disease and epidemiology can be found in several reviews (Gonzalez-Scarano and Nathanson 1996; Schmaljohn 1996; Schmaljohn and Hjelle 1997; Lee et al. 1999; Zaki and Nolte 1999).

3 Hantavirus Pulmonary Syndrome

In 1993, adult respiratory distress (ARDS) cases in the southwestern United States led to the discovery of Sin Nombre virus (SNV) as the etiologic agent of Hantavirus Pulmonary Syndrome (HPS), a new acute respiratory disease (Hjelle et al. 1994; Nolte et al. 1995; Zaki et al. 1995; Zaki and Nolte 1999). Since then, HPS-associated viruses have been identified in at least 28 US States, Canada, and South America. With a 40–45% mortality rate, HPS infections are highly lethal to infected patients. Influenza-like symptoms progress rapidly to acute respiratory distress resulting from bilateral infiltrates within the patients lungs (Nolte et al. 1995; Zaki et al. 1995). The main clinical manifestations of HPS cases are increased vascular permeability, interstitial pneumonia with mononuclear cell infiltrates, severe edema of lung tissue, and acute thrombocytopenia (Nolte et al. 1995; Zaki et al. 1995; Zaki and Nolte 1999). Patients succumb to increased pulmonary edema, which causes pulmonary insufficiency (Nolte et al. 1995; Zaki et al. 1995).

4 Non-Pathogenic Hantaviruses

Although the majority of hantaviruses found to date cause human disease, there are a few hantaviruses which are not associated with pathogenesis. Prospect Hill (PH), Tula (TUL), Thai 749, and Thottapalayam (TPM) have not been associated with any human disease (Schmaljohn and Hjelle 1997). PH virus was first identified in the eastern United States 11 years before the 1993 HPS outbreak in the west, but it appears to be present throughout North America (Yanagihara et al. 1987). There is one report that experimental infection with PH caused a mild transient renal syndrome in lower primates, although no associated disease has been reported in humans (Yanagihara et al. 1988). Tula is present in Europe, and Thai and Thottapalayam are present in southeast Asia, but they have not been studied extensively (Schmaljohn and Hjelle 1997).

5 Common Elements of Hantavirus Disease

Both HFRS and HPS diseases have common elements. Both diseases result in increased vascular permeability and acute thrombocytopenia, although either disease may have pulmonary or renal components. Pathogenic and non-pathogenic hantaviruses all replicate predominantly within pulmonary endothelial cells. However, there is no apparent damage to endothelial cells of humans nor animals (Nolte et al. 1995; Zaki et al. 1995; Lee et al. 1999). Human hantavirus infections are zoonotic, although a report from South America suggests that a case of person-to-person transmission may have occurred with the HPS-associated Andes hantavirus (Lopez et al. 1996; Lopez et al. 1997). However, there have been no prior or additional reports of person-to-person transmission of HFRS- or HPS-causing hantaviruses.

6 Virus–Host Interactions

Hantaviruses are transmitted to humans from small mammal hosts through inhalation of excreted virus (Lee et al. 1978; Schmaljohn 1996). There are a variety of small mammals which serve as hosts for hantaviruses. However, viruses and hosts appear to have co-evolved and specific hantaviruses are primarily associated with a singular species. As a result, the range of the animal host defines the geographic range of the hantavirus and delimits human HFRS or HPS diseases from specific hantaviruses.

- HFRS-causing hantaviruses: Hantaan virus (HTN) is carried by the field mouse, *Apodemus agrarius*, in Asia. Dobrava virus (DOB) is also carried by a discrete species of *Apodemus* (*A. flavicolus*) in the Balkans, Puumala virus (PUU) is carried by bank voles (*C. glariolus*) in Europe, and Seoul virus (SEO) is carried by rats (*R. norvegicus*) worldwide (Schmaljohn and Hjelle 1997).
- HPS-causing hantaviruses: SNV is carried by the deer mouse (*P. maniculatus*), and NY-1 by the white-footed mouse (*P. leucopus*) in North America (Nichol et al. 1993; Hjelle et al. 1995). Many additional genetically and antigenetically distinct hantaviruses are now associated with human HPS in North and South America and are derived from hosts unique to the area of the infection. A complete description of additional HPS hosts and host range can be found in

several articles (Nichol et al. 1993; Elliott et al. 1994; Hjelle et al. 1994; Rollin et al. 1994; Hjelle et al. 1995; Li et al. 1995; Mackow et al. 1995; Ravkov et al. 1995; Schmaljohn et al. 1995; Lopez et al. 1996; Lopez et al. 1997; Schmaljohn and Hjelle 1997; Gavrilovskaya et al. 1999).

- Non-pathogenic hantaviruses: PHV and Tula viruses are carried by species of *Microtus* (*M. pennsylvanicus* and *M. arvalis*) and are respectively present in the United States and Europe. Thai 749 and Thottapalayam are carried by *Bandicota indica* and *Suncus murinus*, respectively (Yanagihara et al. 1987; Arthur et al. 1992; Chu et al. 1994).

In both animals and man hantavirus replication occurs predominantly in pulmonary endothelial cells and macrophages, although viral antigen is present in many organs and prominent in the spleen and kidney as well as the lung (Yanagihara and Silverman 1990; Pensiero et al. 1992; Hughes et al. 1993; Nolte et al. 1995; Zaki et al. 1995; Gonzalez-Scarano and Nathanson 1996; Schmaljohn 1996). However, hantavirus infection of animals does not adversely affect or shorten the host's life span (Gavrilovskaya et al. 1990; Schmaljohn and Hjelle 1997). Hantaviruses are highly viremic for the first month after infection (Gavrilovskaya et al. 1990; Hutchinson et al. 2000). However, animals are persistently infected and still capable of transmitting the virus for over a year. Although serum antibodies appear about 10 days post-infection and result in high-level neutralizing antibody responses, hantaviruses are not cleared from their rodent hosts (Chu et al. 1994). Patients similarly develop neutralizing antibody responses to the virus but are able to clear the virus, suggesting that hantaviruses regulate rodent cellular host responses to effect viral persistence.

Although hantaviruses are carried by a variety of "mice", laboratory mice (*Mus musculus*) are not hantavirus hosts. As an example of the animals that are often termed "mice" that carry hantaviruses, HPS-associated SNV and NY-1 hantaviruses infect *Peromyscus* species, which are more closely related to hamsters or voles than they are to laboratory mice or rats. Although there are varied reports on hantavirus infection of newborn lab mice, there are no reports that *Mus musculus* is a hantavirus host. Newborn lab mice injected with HTN virus reportedly develop a fatal encephalitis and infection of the endothelium (Kim and McKee 1985; McKee et al. 1985; Asada et al. 1987; Tamura et al. 1989). However, 3-week-old mice inoculated with HTN virus develop neither disease nor persistence (Kim and McKee 1985; Asada et al. 1987; Tamura et al. 1989). Exposure of lab mice to highly infected *A. agrarius* hosts carrying HTN also failed to transmit the virus to *M. musculus* (Kim and McKee 1985; McKee et al. 1985; Gonzalez-Scarano and Nathanson 1996). Although there is less known about the transmission of HPS-causing hantaviruses, it is clear that *Mus musculus* are not hosts for HPS-causing hantaviruses.

The lack of disease in hantavirus hosts and the lack of an animal model of hantavirus disease has limited our understanding of hantavirus pathogenesis. The absence of cell lines from host species has also prevented a basic analysis of cellular differences that might contribute to pathogenic and non-pathogenic viral phenotypes. However, the ability of hantaviruses to infect human endothelial cells is a common feature, which permits a focus on hantavirus endothelial cell interactions that may contribute to viral pathogenesis.

7 Cellular Interactions: Keys to Understanding Hantavirus Pathogenesis

Viral attachment and entry into cells is often a major determinant of cellular susceptibility to viral infections. Cellular receptors dictate tissue and cell tropism of viruses and can facilitate viral virulence or host-range restrictions for viral infections (REN et al. 1990; NORKIN 1995). Viral attachment to cells may also stimulate a cascade of intracellular signal transduction events that may activate or regulate cells with pathologic consequences.

The means by which specific hantaviruses cause pulmonary or renal diseases is obscure. Although determinants of pathogenesis have not been defined for any hantavirus, endothelial cells and platelets are prominent regulators of vascular function, and integrins play key roles in barrier functions of these cells (LAMPUGNANI et al. 1991; HAWIGER 1995; TSUKADA et al. 1995; MOGFORD et al. 1996; WU et al. 1998; HODIVALA-DILKE et al. 1999; HYNES et al. 1999; LOSKUTOFF et al. 1999). All hantaviruses infect and replicate in endothelial cells, suggesting that a simple difference in tissue tropism is not a likely determinant of pathogenic or non-pathogenic hantavirus phenotypes (YANAGIHARA and SILVERMAN 1990). This is further suggested by the identical tissue tropism of pathogenic and non-pathogenic hantaviruses within their natural hosts and the lack of disease within animals. However, little is known about hantavirus infection of endothelial cells and immune cells, and their interactions with platelets. These hantavirus cell interactions are likely to be keys to our understanding of: (a) why hantaviruses cause vascular disease; (b) why hantaviruses cause disease in humans but not their animal hosts; (c) why hantaviruses are not cleared from their hosts; (d) what determines host susceptibility; and (e) how hantaviruses establish persistent host infections.

7.1 Endothelial Cell Interactions

All hantaviruses primarily infect pulmonary endothelial cells, although viral antigen is found in endothelial cells in a variety of tissues as well as in immune cells. Our knowledge of hantavirus infections of humans are primarily based on the analysis of pathology specimens and tissue culture interactions with endothelial cells (primarily human umbilical vein endothelial cells, HUVECs) (YANAGIHARA and SILVERMAN 1990; NOLTE et al. 1995; ZAKI et al. 1995; ZAKI and NOLTE 1999). Endothelial cells and platelets regulate vascular permeability by initiating or regulating clotting and complement cascades, or by engaging in a variety of cell–cell interactions which regulate platelet and immune cell recruitment to the vascular endothelium (Cheresh 1987; Jaconi et al. 1991; Hynes 1992; Leavesley et al. 1993; Ylanne et al. 1993; Hemler et al. 1994; Loftus et al. 1994; Shattil et al. 1994; Chang et al. 1995; Salcedo and Patarroyo 1995; Halvorson et al. 1996; Suehiro et al. 1996; Seiffert et al. 1997; Sugimori et al. 1997; Hodivala-Dilke et al. 1999; Hynes et al. 1999; Loskutoff et al. 1999). Several lines of evidence presented below lead to the understanding that cellular integrins are central players in hantavirus–endothelial cell interactions and the differential use of β3 and β1 integrins by pathogenic and non-pathogenic hantaviruses link integrin usage to hantavirus-induced vascular disease (Gavrilovskaya et al. 1998, 1999; Mackow et al. 1999).

Integrins are heterodimeric receptors composed of a combination of α and β subunits that mediate cell–cell adhesion, platelet aggregation, Ca^{+2} channel activation and extracellular matrix (ECM) protein recognition (Cheresh 1987; Phillips et al. 1988; Bossu et al. 1989; Lampugnani et al. 1991; Sepp et al. 1994; Hawiger 1995; Mogford et al. 1996; Mogford et al. 1997; Sugimori et al. 1997; Wu et al. 1998; Hynes et al. 1999). Integrin–ligand interactions mediate the activation and regulation of intracellular signaling pathways that further control both transcriptional and ligand binding functions (Clark et al. 1994; Shattil et al. 1994; Schwartz et al. 1995; Faull et al. 1996). Integrins are also linked to intracellular cytoskeletal elements that facilitate cellular migration on the ECM. ECM proteins such as vitronectin and fibronectin contain Arg-Gly-Asp (RGD) tripeptides, which are recognized by specific cellular integrins including $\alpha_{IIb}\beta_3$, $\alpha_v\beta_3$, and $\alpha_5\beta_1$ (Sugimori et al. 1997). $\alpha_v\beta_3$ and $\alpha_{IIb}\beta_3$ integrins are abundant surface receptors of endothelial cells and platelets, respectively, and $\alpha_v\beta_3$ integrins are present on macrophages (Cheresh 1987; Phillips et al. 1988; Sepp et al. 1994; Hynes et al. 1999; Hynes and Hodivala-Dilke 1999).

7.2 Hantavirus Entry

The mechanism of hantavirus entry into cells has been addressed using inhibitors of endocytosis, lysosomotropic agents, bafilomycin A, and concanamycin (Mackow et al. 1999). Lysosomotropic agents inhibit entry of hantaviruses by 70–80% compared to controls or rhesus rotavirus (RRV), which enters by direct membrane penetration (Gavrilovskaya et al. 1998, 1999; Mackow et al. 1999). Endosome-specific H^+/ATPase inhibitors bafilomycin A and concanamycin (Folimycin) completely block hantavirus entry into cells, demonstrating that endosome acidification is required for viral entry. These findings are consistent with the sensitivity of hantavirus to pH 6 treatment and findings that indicate G1:G2 heterodimers are not formed under these conditions (unpublished observations). Although hantavirus and cell-membrane interactions have not been studied, these results suggest that endosome acidification changes the conformation of G1:G2 heterodimers and permits viral and cellular membrane fusion effecting hantavirus entry.

7.3 Hantavirus Cell Interactions

Initially hantavirus interactions with cell surface components were studied by pretreating cells with enzymes that might abolish viral infectivity. Neuraminidase pretreatment of cells had no effect on hantavirus infectivity, suggesting that hantavirus cell interactions are sialic acid independent (GAVRILOVSKAYA et al. 1998). In contrast, protease pretreatment of cells effectively inhibited hantavirus infectivity, demonstrating that hantaviruses enter endothelial cells by interacting with a proteinaceous cell surface receptor that is internalized through acidified endosomes (GAVRILOVSKAYA et al. 1998; MACKOW et al. 1999).

8 Specific Integrins Mediate Hantavirus Entry into Cells

Cellular receptors for the NY-1 hantavirus were studied by pretreating cells with ligands or antibodies to potential receptors and subsequently analyzing viral entry by immunostaining viral-infected cells 36h post-infection (GAVRILOVSKAYA et al. 1998, 1999; MACKOW et al. 1999). A synopsis of the data, which indicate that β_3 integrins mediate the infection of HPS and HFRS-associated hantaviruses and β_1 integrins mediate the entry of the non-pathogenic PH and Tula hantaviruses, is presented in the following sections (GAVRILOVSKAYA et al. 1998, 1999; MACKOW et al. 1999).

8.1 Integrin Ligands Block Hantavirus Infection

In order to determine whether integrins or additional cell surface receptors mediate the cellular entry of HPS- and HFRS-causing hantaviruses, the ability of ligands to inhibit infection by NY-1, SN, HTN, SEO, PUU, and PH viruses was assessed. A number of ligands to cell surface receptors were tested for their ability to block hantavirus infectivity. Potentially competitive inhibitor proteins, carbohydrates, and lectins were preadsorbed to cells, removed, and cells were then infected to determine the ability of prebound ligands to block virus infectivity. Of the compounds tested, only two ligands had an effect on hantavirus infectivity, vitronectin and fibronectin, and these ligands selectively inhibited pathogenic and non-pathogenic hantaviruses, respectively (GAVRILOVSKAYA et al. 1998, 1999; MACKOW et al. 1999). Pretreatment of cells with vitronectin inhibited the infectivity of pathogenic NY-1, SNV, HTN, SEO, and PUU viruses ($>70\%$) (Fig. 1) (GAVRILOVSKAYA et al. 1998, 1999; MACKOW et al. 1999). In contrast, non-pathogenic PHV, Tula, and TPM viruses were not inhibited by vitronectin but were blocked by fibronectin (Fig. 1). These results suggest that a vitronectin binding cell surface protein serves as a receptor for pathogenic hantaviruses.

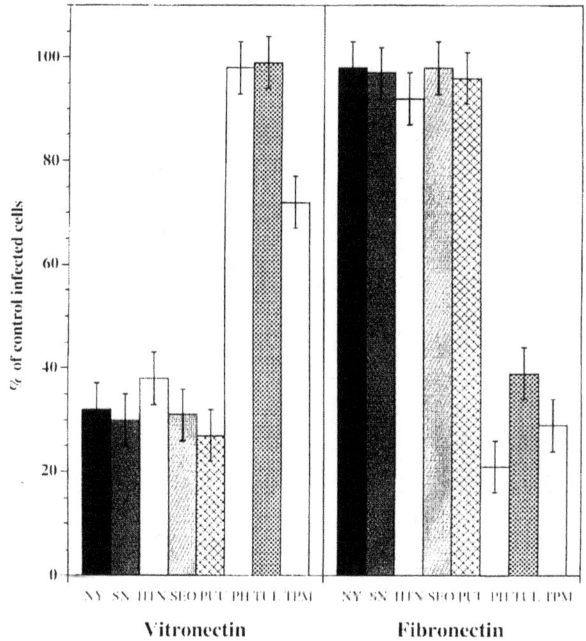

Fig. 1. Ligand-specific inhibition of hantavirus infectivity. Potentially competitive ligands (5–40µg/ml, 50µl) were preadsorbed to Vero E6 cells for 1h before viral adsorption. Approximately 400 FFUs of NY-1, SN, HTN, SEO, PUU, PH, TUL, and TPM hantaviruses were adsorbed to duplicate wells of a 96-well plate. Following adsorption, inocula were removed, and cells were washed and further incubated 24–36h at 37°C in 5% CO_2 before methanol fixation. Hantavirus-infected cells were immunoperoxidase stained as previously described (Gavrilovskaya et al. 1998, 1999; Mackow et al. 1999) using polyclonal rabbit anti-nucleocapsid sera made to bacterially expressed and nickel affinity purified NY-1 N-protein. Infected cells were quantitated and compared to control infections without competitor proteins. Results were reproduced in at least three separate experiments. Results are presented as the percent inhibition of control infections. Pretreatment with 1–500µg/ml BSA did not affect hantavirus infectivity

8.2 Integrin-Specific Antibodies Inhibit Pathogenic and Non-pathogenic Hantaviruses

The ability of the integrin-specific ligands vitronectin and fibronectin to inhibit hantavirus infectivity suggests that integrins facilitate hantavirus entry into cells. In order to directly test this, integrin-specific antibodies, as well as antibodies to other cellular receptors, were tested for their ability to inhibit hantavirus infectivity. Antibodies were prebound to HUVECs or Vero E6 cells. Subsequently, 200 FFUs (focus-forming units) of each hantavirus was adsorbed to cells and infected cells were quantitated 24–36h post-infection. Antibodies to only $\alpha_v\beta_3$ and β_3 were capable of inhibiting infection by pathogenic hantaviruses (SNV, NY-1, HTN, SEO, PUU) (Fig. 2). In contrast, antibodies to additional integrin subunits or other endothelial cell surface proteins did not inhibit infection by pathogenic hantaviruses. Non-pathogenic hantaviruses PHV, Tula, and TPM were not blocked by α_v- or β_3-specific sera and instead were specifically inhibited by sera to α_5 and β_1

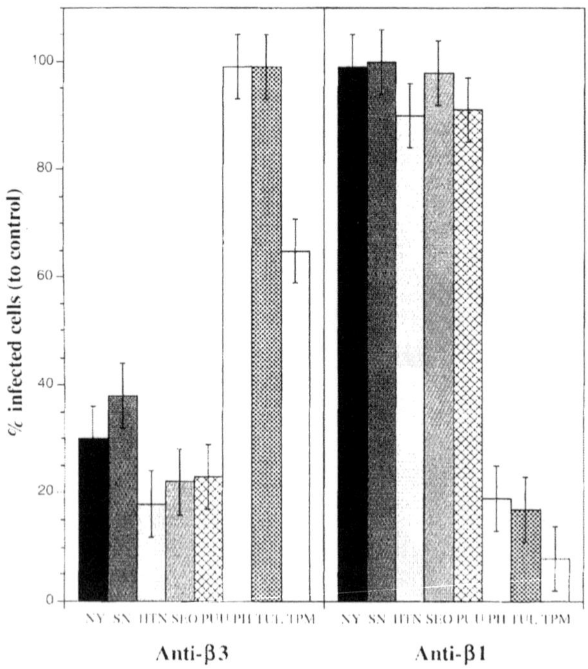

Fig. 2. Infectivity of hantaviruses by integrin-specific antibodies. Duplicate wells of Vero E6 cells were pretreated for 1h (37°C) with 20μg/ml of antibodies to specific integrins. Monolayers were washed and NY-1, SN, HTN, SEO, PUU, PH, TUL, and TPM hantaviruses were adsorbed (Fig. 1) (GAVRILOVSKAYA et al. 1998, 1999; MACKOW et al. 1999). Infected cells were quantitated as in Fig. 1. FFUs observed 36h post-infection are expressed as a percent of control infections for each viral inoculum. Polyclonal rabbit sera to β_1, β_3 were from Chemicon

integrins. $\alpha_v\beta_3$-specific sera inhibited NY-1, SNV, HTN, SEO, and PUU infectivity by approximately 70–90% while failing to block PH infectivity (GAVRILOVSKAYA et al. 1998, 1999; MACKOW et al. 1999). Reciprocally, antisera to $\alpha_5\beta_1$ blocked PH, Tula, and TPM infectivity (>80%) but failed to inhibit NY-1, SNV, HTN, SEO or PUU infections (Fig. 2). These results suggested that β_3-specific integrins mediate the cellular entry of pathogenic hantaviruses, which cause HPS and HFRS. The use of unique integrins by non-pathogenic hantaviruses further suggested that hantavirus pathogenesis may be linked to discrete integrin usage.

8.3 Recombinant Integrins Confer Cell Susceptibility to Pathogenic Hantaviruses

CHO cells lack β_3 integrins and are not permissive for hantavirus infection. In order to demonstrate that β_3 integrins are cellular receptors for pathogenic hantaviruses, we transfected CHO or HEK cells with human recombinant $\alpha_v\beta_3$ or $\alpha_{IIb}\beta_3$ integrins (GAVRILOVSKAYA et al. 1998, 1999; MACKOW et al. 1999). Cell lines

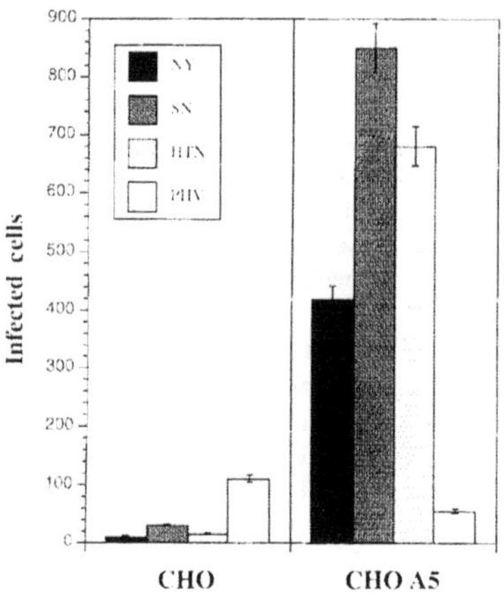

Fig. 3. Recombinant β_3 integrins render CHO cells permissive pathogenic hantaviruses. CHO cells transfected with the $\alpha_{IIb}\beta_3$ integrin (CHO-A5 cells) and infected with NY-1, SN, HTN, and PHV. Following immunoperoxidase staining of nucleocapsid protein in cells, infected cells were quantitated as in Fig. 1 (GAVRILOVSKAYA et al. 1998, 1999; MACKOW et al. 1999)

were tested for their susceptibility to hantavirus infection by quantitating infected cells 24–36h post-infection. CHO or HEK cells containing $\alpha_v\beta_3$ and $\alpha_{IIb}\beta_3$ (CHO A5 cells) integrins were susceptible to SNV, NY-1, HTN, SEO, and PUU virus infections but not to non-pathogenic PHV or Tula hantaviruses (Fig. 3). In addition, pretreatment of CHO cell lines containing recombinant β_3 integrins with β_3 integrin-specific sera inhibited infection by pathogenic hantaviruses. These findings demonstrate that human $\alpha_v\beta_3$ and $\alpha_{IIb}\beta_3$ integrins confer cellular susceptibility to HPS- or HFRS-causing hantaviruses.

8.4 Co-receptors

Regardless of the integrin ligand or antibody blocker employed, some cells are still infected by hantaviruses. Even though some anti-β_3-specific sera block infectivity >95%, the possibility of a hantavirus co-receptor on cells cannot be ruled out. One candidate for a hantavirus co-receptor is the 50-kDa integrin-associated protein (IAP), which is complexed with cellular β_3 integrins. However, IAP-specific antibodies failed to block hantavirus infectivity or to enhance the inhibitory effects of β_3 integrin-specific sera (Fig. 4) (GAVRILOVSKAYA et al. 1999). As a result, it is unlikely that IAPs facilitate hantavirus entry or are required to be complexed with β_3 integrins for entry.

8.5 Human Monoclonal Antibodies to β_3 Integrins Block Hantavirus Infectivity

Humanized antibodies to β_3 integrins are used therapeutically as anti-thrombotic agents. c7E3 (ReoPro) is a commercially available $\alpha_v\beta_3$-specific mouse–human hybrid Fab fragment that is used clinically to inhibit thrombus formation in and around vascular stints (REVERTER et al. 1996; COLLER 1997a,b). Pretreatment of

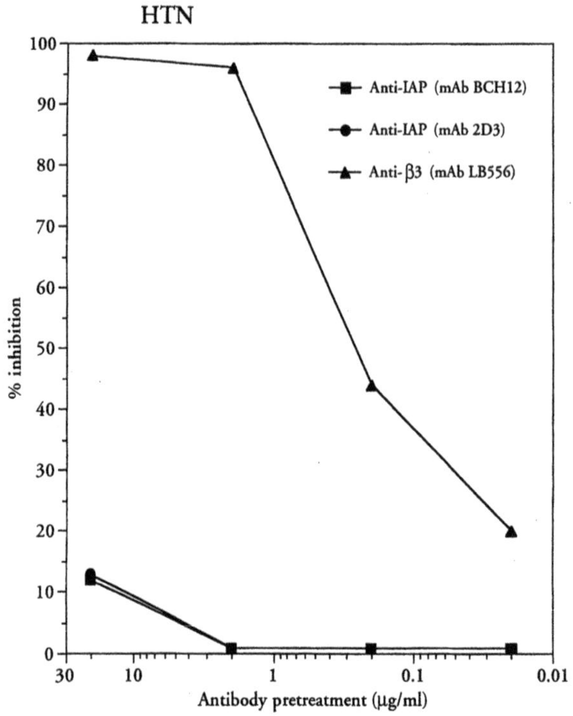

Fig. 4. HTN infectivity is not inhibited antibodies to integrin-associated protein. Duplicate wells of Vero E6 cells were pretreated with 20ng/ml to 20µg/ml monoclonal antibodies, 2D3, and B6H12, to the 50-kDa integrin-associated protein for 1h at 37°C or with mAb LB556, which is specific for β3 integrin subunits (GAVRILOVSKAYA et al. 1998, 1999; MACKOW et al. 1999). Monolayers were washed three times with PBS and infected with approximately 400 FFUs of HTN virus. Infected cells were quantitated as in Fig. 1 and results are presented as the percent inhibition of a control infection

cells with c7E3 or Vitaxin (a distinct hybrid antibody to β_3) reduces the infectivity of SNV, NY-1, HTN, SEO, and PUU viruses by 50–80% while having no effect on the infectivity of PHV or Tula viruses (Fig. 5) (GAVRILOVSKAYA et al. 1999). These findings suggest that anti-β_3 integrin antibodies may be useful therapeutic reagents against hantavirus-induced disease. However, the lack of an animal model of hantavirus disease limits the ability to test the effect of these and other reagents on hantavirus disease processes.

8.6 Hantavirus Interaction with Integrins Is RGD Independent

β_3 subunits confer cell susceptibility to pathogenic hantaviruses, and $\alpha_v\beta_3$ integrins appear to be receptors for pathogenic hantaviruses on endothelial cells. Many integrin ligands bind integrin receptors through RGD tri-peptides, although the specificity of the integrin ligand binding interactions is also mediated by additional, selective interactions (CHERESH 1987; LOFTUS et al. 1990; HUTTENLOCHER et al. 1996; SUEHIRO et al. 1996; TOZER et al. 1996; SUGIMORI et al. 1997; HYNES et al. 1999). However, hantaviruses lack RGD motifs in their G1 and G2 surface glycoproteins, which precludes RGD-specific hantavirus interactions with integrins.

Divalent cations also regulate the high affinity binding state of integrins. Manganese addition enhances high affinity RGD binding by integrins while calcium ion addition inhibits RGD-integrin interactions (SUGIMORI et al. 1997).

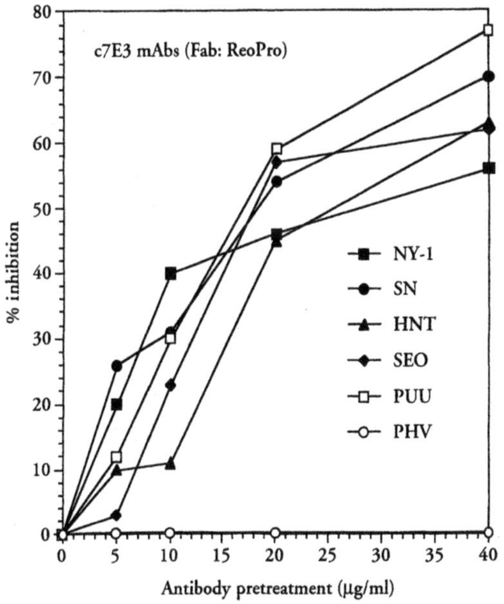

Fig. 5. Fab c7E3 (ReoPro) inhibits the infectivity of HPS- and HFRS-causing hantaviruses. Duplicate wells of Vero E6 cells were pretreated with 20ng/ml to 40μg/ml mouse–human hybrid Fab, c7E3, to $\alpha_v\beta_3$ for 1h at 37°C. Monolayers were washed three times with PBS and infected with approximately 400 FFUs of HTN, SEO, PUU, NY-1, SN, and PH viruses. Infected cells were quantitated as in Fig. 1, and results are presented as the percent inhibition of control infections (GAVRILOVSKAYA et al. 1998, 1999)

Consistent with an RGD independent interaction of hantaviruses with integrins, pretreatment of cells with Mn^{+2} blocks hantavirus infectivity while Ca^{+2} enhances hantavirus infection of cells (GAVRILOVSKAYA et al. 1998, 1999; MACKOW et al. 1999). These results further support the RGD independence of hantavirus integrin interactions and may explain why RGD containing ligands, such as fibronectin, do not block the infectivity of pathogenic hantaviruses (GAVRILOVSKAYA et al. 1998). The lack of an effect of fibronectin on pathogenic hantaviruses may also be explained by the fact that fibronectin is not a high affinity ligand for β_3 integrins (SUGIMORI et al. 1997).

In order to test the RGD independence of hantavirus integrin interactions, experiments were performed with RGD peptides and β_3 integrins, which lack the ability to bind RGD ligands. RGD and RGE peptides were added as competitive inhibitors to determine their ability to block the infection of pathogenic or non-pathogenic hantaviruses (GAVRILOVSKAYA et al. 1998, 1999; MACKOW et al. 1999). At concentrations as high as 200μg/ml, RGD peptides failed to inhibit the infection of any hantavirus tested (NY-1, SNV, HTN, PUU, SEO, PHV, Tula) (GAVRILOVSKAYA et al. 1998, 1999). When a mutant β_3 integrin, which lacks the ability to bind RGD ligands, was transfected into cells, pathogenic hantaviruses were still able to infect cells (GAVRILOVSKAYA et al. 1998, 1999). Neither vitronectin nor RGD peptides had any effect on hantavirus infection of these cells, although β_3-specific sera still blocked infectivity, indicating that the β_3 integrin mutant conferred cell susceptibility to viral infection (GAVRILOVSKAYA et al. 1998). These findings together with the ligand specificity of hantavirus–integrin interactions imply that non-RGD ligand contacts block hantavirus–integrin interactions or that specific integrin vitronectin or integrin fibronectin conformations are unable to interact with hantaviruses.

Interestingly, the $\alpha_v\beta_3$ integrin ligand vitronectin, but not other RGD containing ligands, blocks hantavirus infection. Regions outside of the RGD binding site have been shown to specify binding of fibronectin, fibrinogen, and von Willebrand factor (SUEHIRO et al. 1996; TOZER et al. 1996a,b; SUGIMORI et al. 1997; TOZER et al. 1999). Although this site does not mediate β_3-vitronectin interactions, a ligand-specific interactive region is likely to be present in β_3 integrins, which confers high affinity vitronectin binding by the receptor (TOZER et al. 1996; SUGIMORI et al. 1997). As a result, it is likely that hantaviruses engage a vitronectin binding domain of β_3 integrins or that the dimeric vitronectin sterically blocks hantavirus interactions with β_3 integrins (SMITH and CHERESH 1988; FRELINGER et al. 1990; MARKS et al. 1991; LOSKUTOFF et al. 1999; SCHVARTZ et al. 1999).

8.7 Hantavirus Growth

Hantaviruses grow slowly in tissue culture to maximal titers of approximately 5×10^6FFU/ml (focus forming units per milliliter). Because of the potential for human infection, BSL-3 (biosafety level-3) facilities are required for tissue culture growth of HPS and HFRS strains, and rodent infections require BSL-3 or BSL-4 facilities. Reverse genetic systems have not been established for hantaviruses, and this currently prevents the modification of hantaviruses through recombinant approaches. The lack of reverse genetics, difficulties in studying the heterodimeric G1 and G2 surface glycoproteins of the virus, and the divergence of hantavirus glycoproteins (40–60% divergent) has hampered studies aimed at defining the viral attachment protein.

8.8 Hantavirus Cultivation Does Not Select for Integrin-Specific Viral Variants

Although it is possible that adapting hantaviruses to growth in Vero E6 cells could have selected for hantaviruses that gain entry via β_3 integrins, this is very unlikely. Two reports demonstrate, by passaging hantaviruses in tissue culture, that hantavirus proteins are unaltered (PLYUSNIN et al. 1994; CHIZHIKOV et al. 1995). In one study, the hantavirus genome was sequenced in its entirety from a patient, a small mammal host, and following passage five times in Vero E6 cells, and no amino acid sequence differences were observed in any viral protein (CHIZHIKOV et al. 1995). Although changes were detected in non-coding regions, this demonstrates that viral variants were not selected to use $\alpha_v\beta_3$ integrins by viral adaptation or passage. It has now been demonstrated that five separate hantaviruses, HTN, SEO, PUU, SNV, and NY-1 with up to 60% divergent amino acid sequences in their surface glycoproteins, enter cells via β_3 integrins and are effectively blocked by only one integrin ligand, vitronectin. Similarly, the infectivity of three non-pathogenic hantaviruses is blocked by β_1-specific sera and the β_1 integrin ligand fibronectin

(GAVRILOVSKAYA et al. 1998, 1999; MACKOW et al. 1999). As a result, hantavirus use of β_3 integrins corresponds directly with the pathogenic potential of the virus. The question remaining is how does β_3 integrin usage contribute to hantavirus pathogenesis?

8.9 β_3 Integrins Regulate Capillary Integrity

The potential for pathogenesis to be effected by altering the host state of β_3 integrins is high, and even small differences in the binding affinity or specificity of these receptors may vary the response. β_3 integrins are prominent cell surface receptors on endothelial cells and platelets that mediate platelet activation, endothelial cell adherence, and regulate capillary integrity (CHERESH 1987; HAWIGER 1995; GERBER et al. 1996; HALVORSON et al. 1996; HYNES et al. 1996; SUEHIRO et al. 1996; SUGIMORI et al. 1997; HYNES et al. 1999; HYNES and HODIVALA-DILKE 1999; VEINOT et al. 1999). β_3 integrins include two members, $\alpha_{IIb}\beta_3$ (CD41/CD61) and $\alpha_v\beta_3$, (CD51/CD61) which mediate a diverse group of cellular interactions and signaling responses (CHERESH 1987; HYNES 1992; SHATTIL et al. 1994; SCHWARTZ et al. 1995; HYNES et al. 1999; HYNES and HODIVALA-DILKE 1999). $\alpha_{IIb}\beta_3$ receptors are found on platelets where they mediate and regulate platelet activation and participate in thrombus formation. $\alpha_v\beta_3$ is found on platelets, a number of immune cells, and endothelial cells, and mediates interactions with a variety of cells and ligands (CHARO et al. 1990; SUGIMORI et al. 1997).

β_3 integrins on endothelial cells and platelets play central roles in maintaining capillary integrity (CHERESH 1987; HAWIGER 1995; GERBER et al. 1996; HALVORSON et al. 1996; HYNES and WAGNER 1996; SUEHIRO et al. 1996; SUGIMORI et al. 1997; WU et al. 1998; HYNES et al. 1999; HYNES and HODIVALA-DILKE 1999; VEINOT et al. 1999). Alterations in $\alpha_v\beta_3$ integrin binding permits increased transcapillary fluid fluxes and alter the hydrodynamics of capillaries (HAWIGER 1995; TSUKADA et al. 1995; MOGFORD et al. 1996; HODIVALA-DILKE et al. 1999; HYNES et al. 1999; HYNES and HODIVALA-DILKE 1999). Further, $\alpha_v\beta_3$ and $\alpha_5\beta_1$ differentially regulate arteriolar smooth muscle, resulting in vasodilation or vasoconstriction, respectively, through the intracellular activation of calcium channels (CHARO et al. 1990; HAWIGER 1995; TSUKADA et al. 1995; MOGFORD et al. 1996; SUGIMORI et al. 1997; HODIVALA-DILKE et al. 1999; HYNES et al. 1999; HYNES and HODIVALA-DILKE 1999).

8.10 Role of Vitronectin Interactions in Pathogenesis

$\alpha_v\beta_3$ integrins were originally described as cellular vitronectin receptors (CHARO et al. 1990; SUGIMORI et al. 1997). The use of the vitronectin receptor by hantaviruses and the ability of vitronectin to block hantavirus infectivity also suggests the potential for altered regulation of vitronectin to contribute to viral pathogenesis. Vitronectin is an adhesive glycoprotein present in the blood in an inactive complex with plasminogen activator inhibitor I (PAI-1) (LOSKUTOFF et al. 1999; SCHVARTZ et al. 1999; VEINOT et al. 1999; WOHN et al. 1999). Although vitronectin is present in the blood, the vitronectin RGD is cryptic (SEIFFERT and SMITH 1997) and incapable of binding integrins unless fractionated, activated by PAI-1 in areas of tissue injury or thrombosis, or complexed with thrombin-antithrombin III or complement (C5b–C9) (LOSKUTOFF et al. 1999; SCHVARTZ et al. 1999; WOHN et al. 1999). These interactions result in disulfide-linked multimers of vitronectin and expose heparin binding and RGD binding sites on vitronectin for binding to β_3 integrins (SEIFFERT and SMITH 1997).

The vitronectin-PAI-1 complex stabilizes both PAI-1 and vitronectin and complexed PAI-1 is a thrombin inhibitor (LOSKUTOFF et al. 1999; SCHVARTZ et al. 1999; WOHN et al. 1999). PAI-1 is itself a complex regulatory molecule present in at least five forms that are serine protease inhibitors and anti-adhesive factors in platelet attachment (LOSKUTOFF et al. 1999). Vitronectin binds to PAI, $\alpha_v\beta_3$ integrins, and urokinase plasminogen activator receptor (uPAR) and as a result, it participates in the dynamic regulation of fibrinolysis and cellular adhesion (LOSKUTOFF et al. 1999; SCHVARTZ et al. 1999; WOHN et al. 1999). Vitronectin also inhibits complement mediated cell lysis and competes with heparin binding to antithrombin III, preventing rapid activation of thrombin and thereby functioning in both complement and coagulation systems (LOSKUTOFF et al. 1999). Although it is not clear how these functions are coordinated or regulated, vitronectin provides a unique link between cell adhesion and serum proteolytic activation cascades that could participate in hantavirus-induced thrombocytopenia, increased vascular permeability, or hemorrhage.

8.11 Involvement in Platelet Function

Hantavirus infections of humans result in acute thrombocytopenia. The use of β_3 integrins by hantaviruses and the role of these receptors in platelet activation and recruitment, as well as potential roles for vitronectin and PAI-1 in regulating platelet function, provide a rationale for platelet dysfunction during pathogenic hantavirus infection. Interestingly, humanized monoclonal antibodies, which are used clinically to block platelet aggregation, also block pathogenic hantavirus infectivity. This suggests that hantavirus use of β_3 integrins may dysregulate normal platelet adhesion and activation responses and contribute to thrombocytopenia and vascular permeability changes observed in hantavirus disease.

8.12 Glanzmann's Disease

The involvement of β_3 integrins in vascular disease is prominently documented in a clinical bleeding disorder referred to as Glanzmann's thrombasthenia (BAKER et al. 1997; HYNES et al. 1999; HYNES and HODIVALA-DILKE 1999). Glanzmann's disease is a rare genetic disease of humans characterized by clotting abnormalities and cutaneous hemorrhage (Baker et al. 1997). These deficits result from integrin-specific mutations or deletions within the $\alpha_{IIb}\beta_3$ and $\alpha_v\beta_3$ integrins, which dysregulate platelet aggregation, clot retraction, and endothelial cell adherence properties. Prolonged clotting times, impaired platelet activation, and microvascular bleeding are the primary elements of Glanzmann's disease (Hynes and Hodivala-Dilke 1999). The role of β_3 integrins in Glanzmann's disease suggests a direct means by which viral interaction with β3 integrin receptors may participate in hantavirus pathogenesis.

8.13 β_3-Integrin Knockout Mice Mimic Glanzmann's Disease

A mouse model of human Glanzmann's disease has been developed by generating β_3-integrin subunit knockout mice (murine $\beta_3^-/-$) (Hodivala-Dilke et al. 1999). These mice show nearly identical defects to those of human Glanzmann's disease. Although mice are viable, they have prolonged clotting times and impaired platelet activation, resulting in cutaneous vascular hemorrhage, which leads to reduced survival. These mice demonstrate the direct effect of β_3 integrins on platelet function and vascular permeability, and they illustrate the dramatic role β_3 integrins can play in vascular disease (Hodivala-Dilke et al. 1999). The fact that hantaviruses use β_3 integrins and similar-to-β_3-integrin knockouts, cause thrombocytopenia, and alter vascular permeability (hemorrhage or edema) during infection suggests that β_3-integrin usage is likely to play a central role in the hantavirus disease process.

8.14 β_3 Integrins Contain Species-Specific Differences

The interaction of hantaviruses with integrins from their animal hosts has not been studied, and there is little or no information available about β_3 integrins from these animals. Sequences of β_3 integrin subunits from humans, mice (*Mus musculus*), dogs, pigs, rats, rabbits, and chickens are present in GenBank. Human β_3-integrin sequences differ from these species by ≤15%: rabbit and dog (4%), pig (7%), rat (9%), mouse (10%), and chicken (15%). Human and murine β_3 integrins diverge by 10% at the protein level and include unique residue changes within the metal ion dependent activation site (MIDAS), which could contribute to differential hantavirus interactions (Tozer et al. 1996). Rats are hosts for SEO hantaviruses, yet rat β3 integrin sequences differ from human β3s by nearly as much as the mouse integrin, albeit primarily at unique residues from that of the mouse. An argument can be made that host β3 integrins are not likely to be host receptors for hantaviruses, since no disease results from hantavirus infection of their hosts. However, the evolution of hantaviruses with their hosts may also have selected for non-pathogenic hantavirus-integrin interactions in animal hosts. In fact, the relatively high conservation of β3-integrin subunits, the divergence of hantavirus glycoproteins, and the ability of human β3 integrins to confer cell susceptibility to pathogenic hantaviruses suggests that host β3 integrins are likely to serve as cellular receptors for pathogenic hantaviruses.

8.15 Human, But Not Murine, β_3 Integrins Confer Cell Susceptibility

The inability of hantaviruses to infect adult laboratory mice (*Mus musculus*) and the availability of a recombinant murine β_3 integrin, permitted a test of whether hantavirus host range restriction in mice was at the level of the integrin receptor. CHO cells transfected with plasmids expressing murine or human β3 integrins, in the presence or absence of co-transfected human α_v subunits, were assayed for their susceptibility to hantavirus infection (Fig. 6). Murine β_3 integrin subunits form heterodimers with human α_v subunits that are trafficked to the cell surface. Similarly, human β_3 integrins are transported to the cell surface with the hamster α_v subunit present in CHO cells (Sugimori et al. 1997). The introduction of human β_3 integrins into CHO cells permitted infection by NY-1 or HTN viruses, as previously demonstrated (Gavrilovskaya et al. 1998; Gavrilovskaya et al. 1999; Mackow et al. 1999). Interestingly, cells transfected with the murine β_3 integrin failed to confer cell susceptibility to hantavirus infection (Fig. 6). It has also been shown that transfecting human β_3 integrins into murine cells confers cell susceptibility to hantavirus infection and that human α_v subunits have no effect. These findings suggest that differences between human and murine β_3 integrins determine cellular susceptibility to hantavirus infection and permit the use of recombinant approaches to define domains and residues which are required for infection by pathogenic hantaviruses.

Experiments with recombinant integrins indicate that murine β_3 integrins are insufficient to confer cell susceptibility to pathogenic hantaviruses and that human

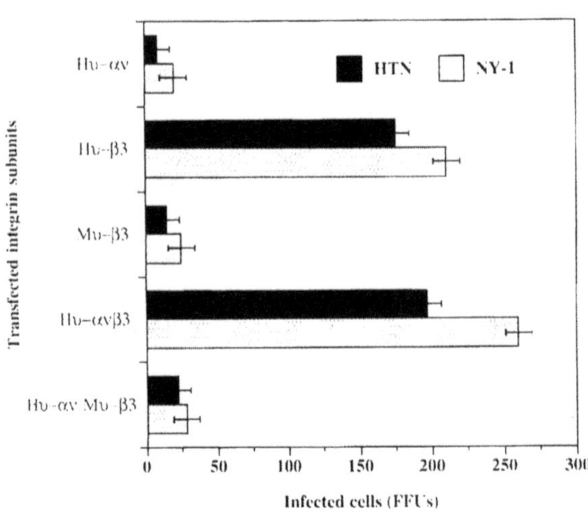

Fig. 6. Human but not Murine β_3 integrins render CHO cells permissive to pathogenic hantaviruses. CHO cells were transfected with human or murine β_3 with and without the human α_v integrin subunits. Transfected cells were infected with NY-1 or HTN 36h post-transfection, and infected cells were quantitated as in Fig. 1 24h post-infection

β_3 integrins supplant murine β_3 deficiencies. These findings suggest: (a) that species specific β_3 integrins are likely to be determinants of hantavirus host range; (b) that human β_3 integrin transgenic mice should confer susceptibility to infection by pathogenic hantaviruses; and (c) that human β_3 integrin transgenic mice have the potential to serve as animal models of hantavirus disease. An analogous situation has been reported for poliovirus, where the human, but not murine, poliovirus receptor confers cell susceptibility, tissue tropism, host range restriction and pathogenesis (MENDELSOHN et al. 1986; MENDELSOHN et al. 1989; RACANIELLO 1991, 1996; MORRISON et al. 1992; REN and RACANIELLO 1992; RACANIELLO et al. 1993; BIBB et al. 1994; RACANIELLO and REN 1994; COLSTON et al. 1995; TAFFS et al. 1996; GROMEIER et al. 1997; SUGIMORI et al. 1997).

8.16 Hantavirus Pathogenesis

The mechanism of hantavirus pathogenesis is still unknown although accumulating data suggest that β_3 integrins are likely to be key elements of hantavirus disease. The means by which some hantaviruses effect hemorrhage or pulmonary edema may stem from virus-specific differences in receptor interactions, alterations in intracellular signaling, the specific induction of cytokines, or the differential regulation of additional platelet or endothelial cell receptors. Large differences in the glycoprotein composition of HFRS- and HPS-causing hantaviruses (approximately 60% and 40% unique residues in G1 and G2 proteins, respectively), further suggest how differential pathogenic responses could be effected by individual viruses despite common β_3 integrin usage. Tumor necrosis factor (TNF) levels in patient sera have been suggested to play a role in hantavirus pathogenesis (MORI et al. 1999) and complement activation has also been suggested to contribute to HFRS disease.

8.17 Lack of Animal Model for Hantavirus Disease

The highly lethal nature of hantavirus infections of humans and the lack of an animal model of hantavirus disease has hampered our ability to study hantavirus pathogenesis. The ability of human β_3 integrins to confer cell susceptibility to hantavirus infection suggests that animal models for hantavirus infection and pathogenesis are plausible. The presence of murine β_3 integrins is likely to ameliorate any disease that might be induced within human β_3 integrin transgenic animals. However, a cross of human β_3 integrin transgenics with murine β_3-integrin knockout mice should both reconstitute β_3 integrin function within the knockouts, by providing the human β_3 subunit, and permit hantavirus infection of murine endothelial cells (HODIVALA-DILKE et al. 1999). If as suggested, the β_3 integrins are primary elements of hantavirus pathogenesis, the human $\beta_3^{+/+}$; murine $\beta_3^{-/-}$ mice are likely be models of hantavirus pathogenesis and permit studies of hantavirus-directed immune responses and therapeutic interventions for hantavirus disease.

8.18 Hantaviruses Regulate β_3 Integrin Function

One of the primary functions of endothelial cells in maintaining capillary integrity and vascular hemostasis is their ability to migrate on extracellular matrix proteins. The ability of β_3 integrins to confer cell susceptibility to hantaviruses and for vitronectin to inhibit hantavirus cell interactions suggested that integrin mediated endothelial cell functions may be regulated by pathogenic hantaviruses. Using FACS analysis the number of β_1 and β_3 integrin receptors on the surface of endothelial cells increased or did not change during infection by NY-1 or PH viruses. However, when the effect of pathogenic (NY-1) and non-pathogenic (PH) hantaviruses on endothelial cell migration was tested, a dramatic difference in cellular response was detected. NY-1 infected cells were able to migrate normally on collagen but migration was inhibited (>60%) on vitronectin (Fig. 7). In contrast, the migration of PH infected cells was increased on collagen and identical to uninfected cells on a vitronectin matrix (Fig. 7). These initial findings suggest that β_3 integrin mediated endothelial cell functions are altered by pathogenic, but not non-pathogenic, hantaviruses and suggest at least one means by which pathogenic hantaviruses may alter vascular permeability during infection.

9 Summary

Hantaviruses cause two potentially lethal diseases, HPS and HFRS, and both diseases result in defects in vascular permeability and platelet function. Human β_3 integrins confer cellular susceptibility to HPS- and HFRS-causing hantaviruses, a fact directly linking platelets, endothelial cells, and hantavirus diseases to the use of

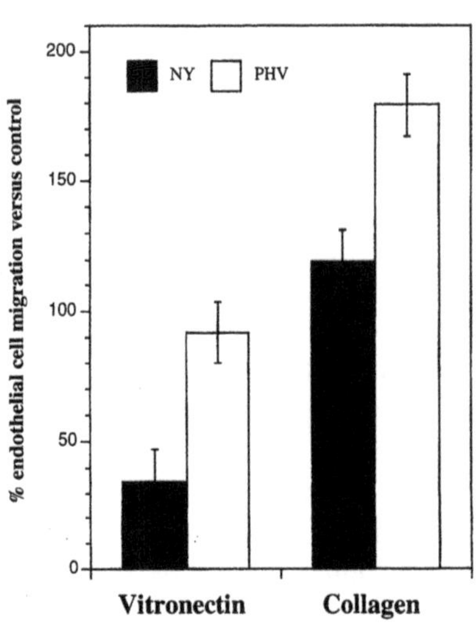

Fig. 7. Hantavirus regulation of endothelial cell migration. Human umbilical vein endothelial cells (HUVECs) were infected at an MOI of 1 with NY-1 or PH viruses. Infected cells were detached from monolayers using enzyme free cell dissociation buffer (GIBCO) and plated on vitronectin or collagen-coated polycarbonate Transwell membranes with 8-μ pores in EBM-2 media (Clonetics). HUVEC migration through pores was quantitated after 3h at 37°C, fixed with DifQuick stain (Dade Bering), and quantitated. Cell migration is presented as a percentage of control uninfected migrated cells. Experiments are representative of three independent experiments

cellular receptors that maintain capillary integrity and regulate platelet function. The role of vitronectin, PAI-1, uPAR, and complement cascades in hantavirus pathogenesis are unstudied but may contribute to specific disease syndromes effected by hantaviruses.

The divergence of hantavirus surface glycoproteins and common β_3-integrin usage provides further insight into the interaction of hantaviruses with cells. G1 and G2 glycoprotein variation is likely to contribute to additional interactions that determine pathogenic responses to individual viruses. β_3-integrin usage also suggests that common elements exist on G1 or the more highly conserved G2 surface glycoprotein, which mediate viral attachment to integrins. Although there is currently no data defining the virion attachment protein, the development of antibodies that recognize the hantavirus attachment protein and block integrin interactions is of interest since it is likely to provide an additional point for therapeutic intervention and vaccine development.

There are a plethora of effects that could be elicited by hantavirus regulation of cellular β_3 integrins and their ligands that are consistent with hantavirus diseases. Since β_3 integrins are critical adhesive receptors on platelets and endothelial cells and regulate both vascular permeability and platelet activation and adhesion, the use of these receptors by hantaviruses is likely to be fundamental to hantavirus pathogenesis. The lack of an animal model for hantavirus pathogenesis has prevented a systematic analysis of immune and cellular responses to hantavirus infections, and it impedes our ability to study protective or therapeutic approaches to hantavirus diseases. However, recent findings suggest that human β_3 integrins within transgenic mice may provide animal models of hantavirus pathogenesis and have the potential to radically alter the ability to investigate hantavirus disease.

Acknowledgements. The authors thank Connie Schmaljohn for providing HTN, SEO, PUU, SN, and PUU viruses, Ric Yanagihara and G.W. Song for providing TUL and TPM viruses, Olli Vapalahti for providing TUL virus, Robert Shope for PH virus, Mark Ginsberg, and Eric Brown for integrin clones and antibodies, and Barry Coller for c7E3 used in these studies. This work was supported by Merit and VA/DOD Awards from the Veterans Administration, and by NIH R03AI42150 and R01AI31016.

References

Arthur RR, Lofts RS, Gomez J, Glass GE, Leduc JW, Childs JE (1992) Grouping of hantaviruses by small (S) genome segment polymerase chain reaction and amplification of viral RNA from wild-caught rats [published erratum appears in Am J Trop Med Hyg 1993 Mar; 48(3):v]. Am J Trop Med Hyg 47(2):210–224

Asada H, Tamura M, Kondo K, Okuno Y, Takahashi Y, Dohi Y, Nagai T, Kurata T, Yamanishi K (1987) Role of T lymphocyte subsets in protection and recovery from Hantaan virus infection in mice. J Gen Virol

Avsic-Zupanc T, Xiao SY, Stojanovic R, Gligic A, van der Groen G, LeDuc JW (1992) Characterization of Dobrava virus: a Hantavirus from Slovenia, Yugoslavia. J Med Virol 38(2):132–137

Baker EK, Tozer EC, Pfaff M, Shattil SJ, Loftus JC, Ginsberg MH (1997) A genetic analysis of integrin function: Glanzmann thrombasthenia in vitro. Proc Natl Acad Sci USA 94(5):1973–1978

Bibb JA, Witherell G, Bernhardt G, Wimmer E (1994) Interaction of poliovirus with its cell surface binding site. Virology 201(1):107–115

Bossu JL, Feltz A, Rodeau JL, Tanzi F (1989) Voltage-dependent transient calcium currents in freshly dissociated capillary endothelial cells. FEBS Lett 255(2):377–380

Chang MC, Jeng JH, Cheong TC, Huang TF (1995) The morphologic change of endothelial cells by ancrod-generated fibrin is triggered by alpha v beta 3 integrin binding and the subsequent activation of a G-protein coupled phospholipase C. Biochimica et Biophysica Acta 1269(2):115–121

Charo I, Nannizzi L, Smith J, Cheresh D (1990) The vitronectin receptor alpha v beta 3 binds fibronectin and acts in concert with alpha 5 beta 1 in promoting cellular attachment and spreading on fibronectin. J Cell Biol 111:2795–2800

Cheresh D (1987) Human endothelial cells synthesize and express an Arg-Gly-Asp-directed adhesion receptor involved in attachment to fibrinogen and von Willebrand factor. Proc Natl Acad Sci 84:6471–6475

Chizhikov V, Spiropoulou C, Morzunov S, Monroe M, Peters C, Nichol S (1995) Complete genetic characterization and analysis of isolation of Sin Nombre Virus. J Virol 69(12):8132–8136

Chu YK, Rossi C, Leduc JW, Lee HW, Schmaljohn CS, Dalrymple JM (1994) Serological relationships among viruses in the Hantavirus genus, family Bunyaviridae. Virology 198(1):196–204

Clark EA, Shattil SJ, Ginsberg MH, Bolen J, Brugge JS (1994) Regulation of the protein tyrosine kinase pp72syk by platelet agonists and the integrin alpha IIb beta 3. J Biol Chem 269(46):28859–28864

Coller B (1997a) GPIIb/IIIa Antagonists: Pathophysiologic and Therapeutic insights from studies of c7E3 Fab. Thrombosis and Haemostasis 78:730–735

Coller B (1997b) Platelet GPIIb/IIIb antagonists: the first anti-integrin receptor therapeutics. J Clin Invest 99:1467–1471

Colston EM, Racaniello VR (1995) Poliovirus variants selected on mutant receptor-expressing cells identify capsid residues that expand receptor recognition. J Virol 69(8):4823–4829

Cosgriff TM, Lewis RM (1991) Mechanisms of disease in hemorrhagic fever with renal syndrome. Kidney Int Suppl 35:S72–S79

Elliott LH, Ksiazek TG, Rollin PE, Spiropoulou CF, Morzunov S, Monroe M, Goldsmith CS, Humphrey CD, Zaki SR, Krebs JW, et al. (1994) Isolation of the causative agent of hantavirus pulmonary syndrome. Am J Trop Med Hyg 51(1):102–108

Faull RJ, Ginsberg MH (1996) Inside-Out Signaling Through Integrins. J Am Soc Neph 7:1091–1097

Frelinger ALD, Cohen I, Plow EF, Smith MA, Roberts J, Lam SC, Ginsberg MH (1990) Selective inhibition of integrin function by antibodies specific for ligand-occupied receptor conformers. J Biol Chem 265(11):6346–6352

Gavrilovskaya I, Apekina N, Bernshtein A, Demina V, Okulova N, Myasnikov Y, Chumakov M (1990) Pathogenesis of hemorrhagic fever with renal syndrome virus infection and mode of horizontal transmission of hantavirus in bank voles. Arch Virol Suppl 1:57–62

Gavrilovskaya I, LaMonica R, Fay ME, Hjelle B, Schmaljohn C, Shaw R, Mackow ER (1999a) New York 1 and Sin Nombre viruses are serotypically distinct viruses associated with hantavirus pulmonary syndrome. J Clin Microbiol 37(1):122–126

Gavrilovskaya I, Podgorodnichenko V, Apekina N, Gordbachkova E, Bogdanova S, Kodkind G, Linev M, Chumakoc M (1987) Determination of Specific Immune complexes and dynamics of their circulation in patients suffering from hemorrhagic fever with a renal syndrome. Microbiol Journal 49:71–76

Gavrilovskaya IN, Brown EJ, Ginsberg MH, Mackow ER (1999b) Cellular entry of hantaviruses which cause hemorrhagic fever with renal syndrome is mediated by beta3 integrins. J Virol 73(5): 3951–3959

Gavrilovskaya IN, Shepley M, Shaw R, Ginsberg MH, Mackow ER (1998) Beta3 Integrins mediate the cellular entry of hantaviruses that cause respiratory failure. Proc Natl Acad Sci USA 95(12): 7074–7079

Gerber DJ, Pereira P, Huang SY, Pelletier C, Tonegawa S (1996) Expression of alpha v and beta 3 integrin chains on murine lymphocytes. Proceedings of the National Academy of Sciences of the United States of America 93(25):14698–14703

Gonzalez-Scarano F, Nathanson N (1996) Bunyaviridae. Virology. Fields. Phila., Lippincott-Ravenpren. 1:1473–1504

Gromeier M, Mueller S, Solecki D, Bossert B, Bernhardt G, Wimmer E (1997) Determinants of poliovirus neurovirulence. J Neurovirol 3 Suppl 1:S35–S38

Halvorson MJ, Coligan JE, Sturmhofel K (1996) The vitronectin receptor (alpha V beta 3) as an example for the role of integrins in T lymphocyte stimulation. Immunologic Research 15(1):16–29

Hawiger J (1995) Mechanisms involved in platelet vessel wall interaction. Thrombosis & Haemostasis 74(1):369–372

Hemler M, Weitzman J, Pasqualini R, Kawaguchi S, Kassner P, Berdichevsky F (1994) Structure, biochemical properties, and biological functions of integrin cytoplasmic domains. In: Takada Y (ed) Integrins: the biological problem. Boca Raton, Fl., CRC Press, pp. 1–36
Hjelle B, Chavez-Giles F, Torrez-Martinez N, Yates T, Sarisky J, Webb J, Ascher M (1994) Genetic identification of a novel hantavirus of the harvest mouse Reithrodontomys megalotis. J Virol 68(10):6751–6754
Hjelle B, Lee SW, Song W, Torrez-Martinez N, Song JW, Yanagihara R, Gavrilovskaya I, Mackow ER (1995) Molecular linkage of hantavirus pulmonary syndrome to the white-footed mouse, Peromyscus leucopus: genetic characterization of the M genome of New York virus. J Virol 69(12):8137–8141
Hjelle B, Spiropoulou CF, Torrez-Martinez N, Morzunov S, Peters CJ, Nichol ST (1994) Detection of Muerto Canyon virus RNA in peripheral blood mononuclear cells from patients with hantavirus pulmonary syndrome. J Infect Dis 170(4):1013–1017
Hodivala-Dilke KM, McHugh KP, Tsakiris DA, Rayburn H, Crowley D, Ullman-Cullere M, Ross FP, Coller BS, Teitelbaum S, Hynes RO (1999) Beta3-integrin-deficient mice are a model for Glanzmann thrombasthenia showing placental defects and reduced survival. J Clin Invest 103(2):229–238
Hughes JM, Peters CJ, Cohen ML, Mahy BW (1993) Hantavirus pulmonary syndrome: an emerging infectious disease. Science 262(5135):850–851
Hutchinson KL, Rollin PE, Shieh WJ, Zaki S, Greer PW, Peters CJ (2000) Transmission of Black Creek Canal virus between cotton rats. J Med Virol 60(1):70–76
Huttenlocher A, Ginsberg MH, Horwitz AF (1996) Modulation of cell migration by integrin-mediated cytoskeletal linkages and ligand-binding affinity. J Cell Biol 134(6):1551–1562
Hynes R (1992) Integrins: Versatility, modulation and signaling in cell adhesion. Cell 69:11–25
Hynes RO, Bader BL, Hodivala-Dilke K (1999) Integrins in vascular development. Braz J Med Biol Res 32(5):501–510
Hynes RO, Hodivala-Dilke KM (1999) Insights and Questions arising from the studies of a mouse model of Glanzmann Thrombasthenia. Thrombosis and Haemostasis 82:481–485
Hynes RO, Wagner DD (1996) Genetic manipulation of vascular adhesion molecules in mice. J Clin Invest 98(10):2193–2195
Jaconi M, Theler J, Schlegel W, Appel R, Wright S, Lew P (1991) Multiple elevations of cytosolic free $Ca2^{+}$ in human neutrophils: initiation by adherence receptors of the integrin family. J Cell Biol 112:1249–1257
Kim GR, McKee KJ (1985) Pathogenesis of Hantaan virus infection in suckling mice: clinical, virologic, and serologic observations. Am J Trop Med Hyg 34(2):388–395
Lampugnani M, Resnati M, Dejana E, Marchisiio P (1991) The role of integrins in maintenance of endothelial monolayer integrity. J Cell Biol 112:479–490
Leavesley D, Schwartz M, Rosenfeld M, Cheresh D (1993) Integrin b1 and b3-mediated endothelial cell migration is triggered through distinct signaling mechanisms. J Cell Biol 121:163–170
Lee H, Lee P, Johnson K (1978) Isolation of the etiologic agent of Korean hemorrhagic fever. J Infect Dis 137:298–308
Lee HW (1982) Hemorrhagic fever with renal syndrome (HFRS) Scand J Infect Dis Suppl 36(82):82–85
Lee JS, Lahdevirta J, Koster F, Levy H (1999) Clinical Manifestations and Treatment of HFRS and HPS. In: Lee HW, Calisher C, Schmaljohn C (eds.) Manual of Hemorrhagic Fever with Renal Syndrome and Hantavirus Pulmonary Syndrome. Seoul, Asan Institute for Life Sciences, pp 17–38
Li D, Schmaljohn AL, Anderson K, Schmaljohn CS (1995) Complete nucleotide sequences of the M and S segments of two hantavirus isolates from California: evidence for reassortment in nature among viruses related to hantavirus pulmonary syndrome. Virology 206:973–983
Loftus J, O'Toole T, Plow E, Glass A, Frelinger III A, Ginsberg M (1990) A B3 integrin mutation abolishes ligand binding and alters divalent cation-dependent conformation. Science 249:915–918
Loftus J, Smith J, Ginsberg M (1994) Integrin-mediated cell adhesion: the extracellular face. J Biol Chem 269:25235–25238
Lopez N, Padula P, Rossi C, Lazaro ME, Franze-Fernandez MT (1996) Genetic identification of a new hantavirus causing severe pulmonary syndrome in Argentina. Virology 220(1):223–226
Lopez N, Padula P, Rossi C, Miguel S, Edelstein A, Ramirez E, Franze-Fernandez MT (1997) Genetic characterization and phylogeny of Andes virus and variants from Argentina and Chile. Virus Res 50(1):77–84
Loskutoff DJ, Curriden SA, Hu G, Deng G (1999) Regulation of cell adhesion by PAI-1. Apmis 107(1):54–61
Mackow E, Luft B, Bosler E, Goldgaber D, Gavrilovskaya I (1995) Hantavirus pulmonary syndrome viruses identified on Shelter Island, NY. New England Journal Of Medicine 332(5):337–338

Mackow ER, Ginsberg MH, Gavrilovskaya IN (1999) Beta3 integrins mediate the cellular entry of pathogenic hantaviruses. In: Saluzzo JF, Dodet B (eds) Factors in the emergence and control of rodent-borne viral diseases (Hantaviral and Arenal Diseases). Paris, Elsevier, pp 113–123
Marks P, Hendey B, Maxfield F (1991) Attachment to Fibronectin or Vitronectin makes human neutrophil migration sensitive to alterations in cytosolic free calcium concentration. J Cell Biol 112:149–158
McKee KJ, Kim GR, Green DE, Peters CJ (1985) Hantaan virus infection in suckling mice: virologic and pathologic correlates. J Med Virol 17(2):107–117
Mendelsohn C, Johnson B, Lionetti KA, Nobis P, Wimmer E, Racaniello VR (1986) Transformation of a human poliovirus receptor gene into mouse cells. Proc Natl Acad Sci USA 83(20):7845–7849
Mendelsohn CL, Wimmer E, Racaniello VR (1989) Cellular receptor for poliovirus: molecular cloning, nucleotide sequence, and expression of a new member of the immunoglobulin superfamily. Cell 56(5):855–865
Mogford JE, Davis GE, Meininger GA (1997) RGDN peptide interaction with endothelial alpha5beta1 integrin causes sustained endothelin-dependent vasoconstriction of rat skeletal muscle arterioles. J Clin Invest 100(6):1647–1653
Mogford JE, Davis GE, Platts SH, Meininger GA (1996) Vascular smooth muscle alpha v beta 3 integrin mediates arteriolar vasodilation in response to RGD peptides. Circ Res 79(4):821–826
Mori M, Rothman AL, Kurane I, Montoya JM, Nolte KB, Norman JE, Waite DC, Koster FT, Ennis FA (1999) High levels of cytokine-producing cells in the lung tissues of patients with fatal hantavirus pulmonary syndrome. J Infect Dis 179(2):295–302
Morrison ME, Racaniello VR (1992) Molecular cloning and expression of a murine homolog of the human poliovirus receptor gene. J Virol 66(5):2807–2813
Nichol ST, Spiropoulou CF, Morzunov S, Rollin PE, Ksiazek TG, Feldmann H, Sanchez A, Childs J, Zaki S, Peters CJ (1993) Genetic identification of a hantavirus associated with an outbreak of acute respiratory illness [see comments]. Science 262(5135):914–917
Nolte KB, Feddersen RM, Foucar K, Zaki SR, Koster FT, Madar D, Merlin TL, McFeeley PJ, Umland ET, Zumwalt RE (1995) Hantavirus pulmonary syndrome in the United States: a pathological description of a disease caused by a new agent. Human Pathology 26(1):110–120
Norkin LC (1995) Virus receptors: implications for pathogenesis and the design of antiviral agents. Clinical Microbiology Reviews 8(2):293–315
Pensiero MN, Sharefkin JB, Dieffenbach CW, Hay J (1992) Hantaan virus infection of human endothelial cells. J Virol 66(10):5929–5936
Phillips D, Charo I, Parise L (1988) The platelet membrane glycoprotein GPIIb/IIIa complex. Blood 71:831–43
Plyusnin A, Vapalahti O, Ulfves K, Lehvaslaiho H, Apekina N, Gavrilovskaya I, Blinov V, Vaheri A (1994) Sequences of wild Puumala virus genes show a correlation of genetic variation with geographic origin of the strains. J Gen Virol 75(Pt 2):405–409
Racaniello VR (1991) Virus-receptor interaction in poliovirus entry and pathogenesis. Harvey Lect 87:1–16
Racaniello VR (1996) Early events in poliovirus infection: virus-receptor interactions. Proc Natl Acad Sci USA 93(21):11378–11381
Racaniello VR, Ren R (1994) Transgenic mice and the pathogenesis of poliomyelitis. Arch Virol Suppl 9:79–86
Racaniello VR, Ren R, Bouchard M (1993) Poliovirus attenuation and pathogenesis in a transgenic mouse model for poliomyelitis. Dev Biol Stand 78:109–116
Ravkov EV, Rollin PE, Ksiazek TG, Peters CJ, Nichol ST (1995) Genetic and serologic analysis of Black Creek Canal virus and its association with human disease and Sigmodon hispidus infection. Virology 210(2):482–489
Ren R, Racaniello VR (1992) Human poliovirus receptor gene expression and poliovirus tissue tropism in transgenic mice. J Virol 66(1):296–304
Ren RB, Costantini F, Gorgacz EJ, Lee JJ, Racaniello VR (1990) Transgenic mice expressing a human poliovirus receptor: a new model for poliomyelitis. Cell 63(2):353–362
Reverter J, Beguin S, Kessels H, Kumar R, Hemler H, Coller B (1996) Inhibition of Platelet-mediated, tissue factor-induced thrombin generation by the Mouse/human chimeric 7E3 antibody. J Clin Invest 98(3):863–874
Rollin PE, Coudrier D, Sureau P (1994) Hantavirus epidemic in Europe, 1993 [letter] [see comments]. Lancet 343(8889):115–116

Schmaljohn AL, Li D, Negley DL, Bressler DS, Turell MJ, Korch GW, Ascher MS, Schmaljohn CS (1995) Isolation and initial characterization of a newfound hantavirus from California. Virology 206(2):963–972

Schmaljohn C (1996) Bunyaviridae and their replication. Fields Virology. B. Fields. Phila., Lippencott-Raven Press. 1:1447–1471

Schmaljohn C, Hjelle B (1997) Hantaviruses: a global disease problem. Emerg Infect Dis 3(2):95–104

Schvartz I, Seger D, Shaltiel S (1999) Vitronectin. Int J Biochem Cell Biol 31(5):539–544

Schwartz M, Shaller M, Ginsberg M (1995) Integrins: emerging paradigms of signal transduction. Annu Rev Cell Dev Biol 11:549–599

Seiffert D, Smith JW (1997) The cell adhesion domain in plasma vitronectin is cryptic. J Biol Chem 272(21):13705–13710

Sepp NT, Li LJ, Lee KH, Brown EJ, Caughman SW, Lawley TJ, Swerlick RA (1994) Basic fibroblast growth factor increases expression of the alpha v beta 3 integrin complex on human microvascular endothelial cells. J Invest Dermatology 103(3):295–299

Shattil SJ, Ginsberg MH, Brugge JS (1994) Adhesive signaling in platelets. Curr Opin Cell Biol 6(5):695–704

Smith J, Cheresh D (1988) The Arg-Gly-Asp binding domain of the vitronectin receptor. J Biol Chem 263:18726–18731

Suehiro K, Smoth J, Plow E (1996) The Ligand Specifcity of β3 Integrins. J Biol Chem 271(17):10365–10371

Sugimori T, Griffith D, Arnaout M (1997) Emerging Paradigms of Integrin Ligand Binding and Activation. Kidney International 51:1454–1462

Taffs R, Dragunsky E, Nomura T, Hioki K, Levenbook I, Fitzgerald E (1996) Importance of experiments in poliovirus-susceptible transgenic mice for evaluating current potency tests of inactivated polio vaccine (IPV) Dev Biol Stand 86:345

Tamura M, Asada H, Kondo K, Tanishita O, Kurata T, Yamanishi K (1989) Pathogenesis of Hantaan virus in mice. J Gen Virol

Tozer EC, Baker EK, Ginsberg MH, Loftus JC (1999) A mutation in the alpha subunit of the platelet integrin alphaIIbbeta3 identifies a novel region important for ligand binding. Blood 93(3):918–24

Tozer EC, Hughes PE, Loftus JC (1996a) Ligand binding and affinity modulation of integrins. Biochem Cell Biol 74(6):785–798

Tozer EC, Liddington RC, Sutcliffe MJ, Smeeton AH, Loftus JC (1996b) Ligand binding to integrin alphaIIbbeta3 is dependent on a MIDAS-like domain in the beta3 subunit. J Biol Chem 271(36):21978–21984

Tsukada H, Ying X, Fu C, Ishikawa S, McKeown-Longo P, Albelda S, Bhattacharya S, Bray BA, Bhattacharya J (1995) Ligation of endothelial alpha v beta 3 integrin increases capillary hydraulic conductivity of rat lung. Circulation Research 77(4):651–659

Veinot JP, Srivatsa S, Carlson P (1999) Beta3 integrin – a promiscuous integrin involved in vascular pathology. Can J Cardiol 15(7):762–770

Wohn KD, Schmidt T, Kanse SM, Yutzy B, Germer M, Morgenstern E, Preissner KT (1999) The role of plasminogen activator inhibitor-1 as inhibitor of platelet and megakaryoblastic cell adhesion. Br J Haematol 104(4):901–908

Wu X, Mogford JE, Platts SH, Davis GE, Meininger GA, Davis MJ (1998) Modulation of calcium current in arteriolar smooth muscle by alphavbeta3 and alpha5beta1 integrin ligands [in process citation]. J Cell Biol 143(1):241–252

Yanagihara R, Amyx HL, Lee PW, Asher DM, Gibbs CJ, Gajdusek DC (1988) Experimental hantavirus infection in nonhuman primates. Arch Virol 101(1–2):125–130

Yanagihara R, Daum CA, Lee PW, Baek LJ, Amyx HL, Gajdusek DC, Gibbs CJ (1987) Serological survey of Prospect Hill virus infection in indigenous wild rodents in the USA. Trans R Soc Trop Med Hyg 81(1):42–45

Yanagihara R, Silverman DJ (1990) Experimental infection of human vascular endothelial cells by pathogenic and nonpathogenic hantaviruses. Arch Virol 111(3–4):281–286

Ylanne J, Chen Y, O'Toole TE, Loftus JC, Takada Y, Ginsberg MH (1993) Distinct functions of integrin alpha and beta subunit cytoplasmic domains in cell spreading and formation of focal adhesions. J Cell Biol 122(1):223–233

Zaki S, Greer P, Coffield L, Goldsmith C, Nolte K, Foucar K, Feddersen R, Zumwalt R, Miller G, Rollin P, et al. (1995) Hantavirus pulmonary syndrome: pathogenesis of an emerging infectious disease. Am J Pathol 146:552–579

Zaki SR, Nolte KB (1999) Pathology, Immunohistochemistry and in situ hybridization. In: Lee HW, Calisher C, Schmaljohn C (eds) Manual of hemorrhagic fever with renal syndrome and hantavirus pulmonary syndrome. Seoul, ASAN Institute for Life Sciences,143–154

Clinical Manifestations of New World Hantaviruses

D.A. Enria, A.M. Briggiler, N. Pini, and S. Levis

1 Introduction

In May 1993 a new clinical entity named hantavirus pulmonary syndrome (HPS) was first described in the United States during the investigation of an outbreak of a febrile disease of high mortality, characterized by acute respiratory insufficiency (Duchin et al. 1994). HPS was found to be caused by a novel agent of the *Hantavirus* genus, Sin Nombre virus (SNV; Nichol et al. 1993; Ksiazek et al.

Instituto Nacional de Enfermedades Virales Humanas "Dr. Julio I. Maiztegui" (INEVH), ANLIS, Monteagudo 2510, 2700 Pergamino, Argentina

1995). Since the discovery of the disease, isolated cases and outbreaks of HPS have been registered all over the Americas, and other SNV-related hantaviruses (New World hantaviruses) were identified either causing human illnesses or circulating among native sigmodontine rodent species.

In this chapter, we will discuss the clinical manifestations of New World hantaviruses, with an attempt to analyze the differences that have been observed among different geographic areas and viral genotypes.

2 History of Outbreaks and Discovery of Agents

Hantaviruses represent a unique genus of the *Bunyaviridae* family because all their members (except for Thottapalayam virus) are rodent-borne, while the other viruses belonging to four genera are arthropod-borne. They are lipid-enveloped, spherical viruses consisting of an RNA trisegmented genome (Schmaljohn and Hjelle 1997).

Until 1993, human illnesses caused by hantaviruses were grouped under the name of hemorrhagic fever with renal syndrome (HFRS), and their distribution was limited to the Eurasian landmark. Known hantaviruses causing HFRS included Hantaan (prototype member of the family), Puumala, Dobrava, and Seoul viruses.

In America, Prospect Hill was the first hantavirus isolated from a rodent source, with no associated illness described. The probability of Seoul virus' worldwide distribution in relation to the cosmopolitan reservoir, *Rattus norvegicus*, prompted researchers to look for activity of hantaviruses in rodent reservoirs as well as for human infections. These illnesses were originally sought under the clinical definition of HFRS. Several studies performed in the 1980s showed data of activity of hantaviruses among wild and laboratory rodents, and provided evidences of human infections in North and South America (Maiztegui et al. 1983; LeDuc et al. 1985, 1986; Weissenbacher et al. 1990, 1996; Parisi et al. 1996).

After the discovery of SNV, HPS was diagnosed in another seven countries of the Americas (Argentina, Bolivia, Brazil, Canada, Chile, Paraguay, and Uruguay), and other SNV-related members were identified from indigenous rodent species (Iversson et al. 1994; Khan et al. 1995, 1996a; Lopez et al. 1996; Johnson et al. 1997; Levis et al. 1998). New World hantaviruses associated with human illnesses so far are: New York, Bayou, Black Creek Canal, Juquitiba, Laguna Negra, Orán, Hu39694, Lechiguanas, and Andes (Fig. 1).

Bayou, Black Creek Canal, and New York viruses were discovered from the study of suspected HPS cases from outside the initially 1993-involved states in the United States (Peters 1998). Juquitiba was identified after an outbreak of HPS in 1993, affecting three brothers living in the rural area of Sao Paulo State, Brazil. Laguna Negra was discovered through an investigation of an outbreak of HPS that occurred on an agricultural community in the western Chaco region of Paraguay in 1995.

In Argentina, four different genotypes causing human illness have been described so far. Orán is the main genotype causing HPS in northern Argentina.

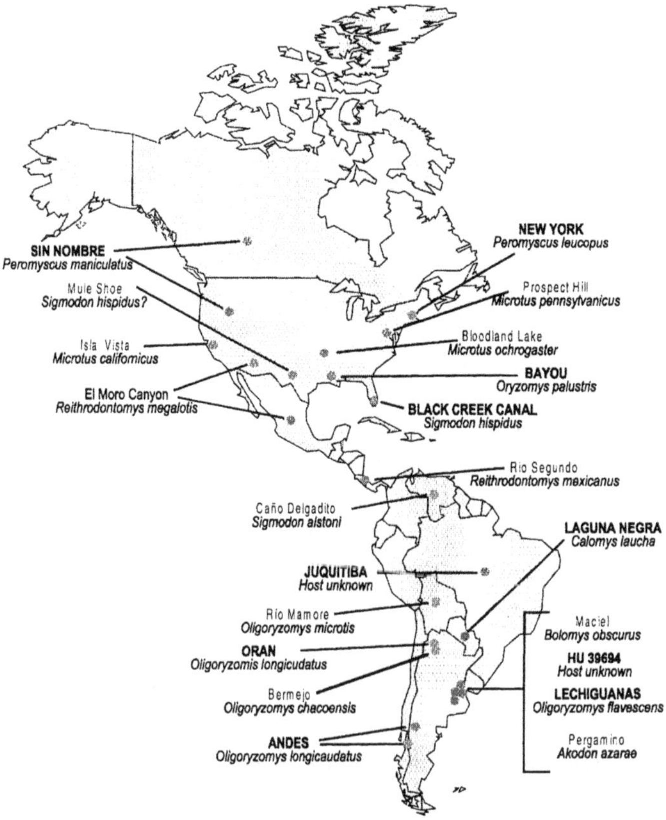

Fig. 1. New World hantaviruses. For each member, putative reservoir is given. New World hantaviruses associated with human illnesses to date (*bold*)

Interestingly, in the 1980s in this region there had been reports of an unusual presentation of an acute respiratory distress syndrome of unknown etiology. The syndrome was known as "Distress of Orán". All the studies to establish etiology were unsuccessful up to the early 1990s, when some of these cases were found to be caused by *Leptospira interrogans*. After 1993, it was found that another group of cases with Distress of Orán were etiologically associated with an SNV-related hantavirus, ultimately referred as Orán. Lechiguanas and Hu39694 are the known viral genotypes causing human diseases in Central Argentina. They were discovered when investigators were looking for (a) hantavirus-associated illnesses among Argentine hemorrhagic fever (an hemorrhagic fever caused by the arenavirus Junin), which tested negative for laboratory evidences of acute Junin virus infection; (b) leptospirosis-suspected cases that tested negative for laboratory evidences of Leptospira infections; and (c) after 1993, cases fulfilling the HPS case definition according to Centers for Disease Control and Prevention (CDC) outlines (CDC 1997). Andes virus was discovered as the causal agent of an HPS outbreak occurring in southern Argentina in 1995, and was subsequently shown to be the virus related to most of the cases registered in Chile (LOPEZ et al. 1996; LOPEZ et al. 1997; ENRIA 1998).

3 Clinical Manifestations of Infections Caused by SNV

The original description of HPS as a new clinical entity caused by hantaviruses came from the study of SNV human infections. SNV is the major agent responsible for HPS in North America (DUCHIN et al. 1994; KOSTER and LEVY 1998; SIMPSON and LEVY 1999).

The disease is characterized by an incubation period ranging from 1 to 5 weeks. In some cases in which the information available allows a better estimation, the period is considered to be between 9 and 24 days.

Clinical features of HPS were originally divided into four phases (febrile, cardiopulmonary, diuretic, and convalescent), in contrast to the five classically known for HFRS.

3.1 Febrile Phase

This prodromal period, typically lasting from 3 to 5 days (range 1–12 days) is characterized by fever, myalgia, and malaise. Other symptoms that may be present include headache, dizziness, anorexia, nausea, vomiting, and diarrhea. In some patients, there is abdominal pain that could be confused with appendicitis or pyelonephritis. Early recognition of HPS cases in this period is difficult, and the disease is indistinguishable from other viral prodromes. In this respect, it is important to consider that signs of upper respiratory tract disease such as sore throat, rhinorrhea, sinusitis, and ear pain are commonly not present. At the end of this prodromal phase, and announcing the onset of pulmonary edema, non-productive cough and tachypnea appear. On physical examination, rales or signs of pleural effusion may sometimes be present.

3.2 Cardiopulmonary Phase

This phase is characterized by the presentation of shock and pulmonary edema, that may progress rapidly (in 4–24h). Hypotension and oliguria, sometimes with weakness and delirium, are the result of shock. Tachypnea, exertional dyspnea, and non-productive cough are the expression of pulmonary edema. This edema is non-cardiogenic in origin, and pulmonary capillary wedge pressures obtained through a Swan-Ganz catheter are normal, as is the heart size on the X-ray. Hypoxemia is a common feature, with an oxygen saturation of hemoglobin less than 95% at sea level. The progressive leakage of high protein fluid from circulating blood into the lung interstitium and alveoli results in hypovolemia, which contributes to shock. Myocardial index sometimes less than 2.2l/min/m^2 is indicative of myocardial failure, further contributing to shock. Once pulmonary edema is present, the disease proceeds fast, with patients dying within 24–48h, hypoxia, circulatory compromise, or both being the immediate cause of death. The circulatory collapse at this stage differs from the finding in septic shock, and the most commonly registered pattern seen is low cardiac output, high systemic vascular resistance, and hypotension.

3.3 Diuretic Phase

This phase is characterized by a rapid clearance of pulmonary edema, resolution of fever and shock. Spontaneous diuresis is an early sign of this process.

3.4 Convalescent Phase

This period may last up to 2 months, with patients recovering apparently fully. However, preliminary evidences suggest that pulmonary dysfunction, illustrated by decreased small-airways flow and diminished diffusing capacity, may be a common finding in the short term, but a continued follow-up is still necessary to determine the long-term persistence of these alterations.

4 Clinical Laboratory Findings

In SNV infection, thrombocytopenia with platelet counts less than 150,000/mm^3 is almost constant (98% of cases), although it may be not present in early stages. The white cell count can be normal or elevated on admission and frequently increases to very high values. A left shift and the presence of immunoblasts (prominent mononuclear cells of more than 180μ in diameter, with deep blue cytoplasm by Giemsa stain) is frequently seen, but may not appear until late in the prodromal phase, or even when pulmonary edema is already installed. Hemoconcentration is also a frequent finding (up to 81% of cases) (Khan et al. 1996b).

The kidney is not usually compromised in a major way as in HFRS, and elevated creatinine and blood urea nitrogen are in relation to shock and hypovolemia. Proteinuria and abnormal urinary sediment are often registered.

Other findings include elevated hepatic enzymes (rarely exceeding five times the upper limit), with normal bilirubin levels; hypoalbuminemia, metabolic acidosis, slightly elevated aspartate aminotransferase, elevated lactate dehydrogenase, and in severe cases lactic acidosis. Although usually subclinical, most patients exhibit evidence of a coagulopathy that, apart from the thrombocytopenia, includes an abnormal partial thromboplastin time, and sometimes a prolonged prothrombin time. Circulating D-dimers are uncommon, as are fibrinogen levels falling below 200mg/dl.

5 Radiological Findings

Radiography of the chest is also useful for the diagnosis of HPS. Although normal findings or asymmetrical compromise may occur sporadically in the early stages, progression from slight interstitial edema (with Kerley B lines, peribronchial cuffing, and alveolar interstitial fluid in the basal segments) to bilateral alveolar edema is very characteristic. Pleural effusion is also frequently evident (Ketai et al. 1994).

6 Reported Clinical and Epidemiological Differences Among New World Hantavirus Infections

Some differences from the classical SNV infections have been reported in other regions of the Americas or have been associated with other New World hantaviruses. These include the possibility of person-to-person transmission, the occurrence of childhood disease, the presence of high seropositivity rates in some areas of South America, the spectrum of disease severity, the degree of renal compromise, and hemorrhagic signs (Table 1).

6.1 Person-to-Person Transmission

Until 1996, interhuman transmission of hantavirus had never been documented, with strong evidence in this respect from Europe and Asia for Old World hantaviruses, and the United States for SNV. In March 1995, a familiar outbreak of HPS in El Bolsón, southern Argentina, led to the discovery of Andes virus. In an unusual HPS outbreak due to Andes virus in the same area in 1996, a chain of apparent interhuman transmission was registered in 16 out of 20 cases, with a strong suggestion that health personnel could acquire the infection from patients with HPS. In all of these cases except one, infection from rodents could not be ruled out because the patients had lived in or had visited the endemic area in a period that was coincident with the incubation period of the disease. The exception that gave strong evidence for the occurrence of interhuman transmission was the case of a

Table 1. Reported clinical and epidemiological differences among New World Hantavirus infections

	Differences among regions/genotypes
Spectrum of clinical illnesses	Mild disease: rare for SNV; 11–13% in Argentina/Chile; probably high in Paraguay
Prevalence rates	United States: 0.3% overall
	Argentina, Central and South: 0.1–1.5%; Northern Argentina: 18–20%
	Chile: 2–13%
	Paraguay: 7–21%; 40% in Gran Chaco Indians
Interhuman transmission	Only demonstrated for Andes virus in 1996 outbreak in El Bolsón (Argentina)
Infection in children (less than age 17)	United States: 4.5–6.9% – Youngest infected case: 4 years old
	Argentina: 12–14.6% – Youngest infected case: 1 year old
	Chile: 15.2% in 1997 outbreak
Renal compromise	SNV: rare; Bayou: 3/3; Black Creek Canal: 1/1
	Andes: ~50%; Lechiguanas: ~50%; Orán probably ~50%
Hemorrhagic manifestations	Apparently more frequent for Andes virus infections

doctor from Buenos Aires (Argentina's capital city), that had never visited the outbreak area, and whose unique risk factor was to have treated an HPS case that had traveled from El Bolsón to Buenos Aires. Case-control studies identified contact with other cases as a major risk factor; rodents studies showed population densities in the outbreak area lower than expected, and molecular studies established a close similarity of the viral genome in the cases studied (Enria et al. 1996; Wells et al. 1997a; Padula et al. 1998).

So far, this has been the only situation in which person-to-person transmission of Andes virus was demonstrated. Seroprevalence in a random household survey in southern Argentina was 1%, and no antibodies were found in a survey of 150 healthcare workers from the outbreak region (Wells et al. 1998). In other familial clusters of cases associated with Andes virus and occurring in both Chile and Argentina, the situation has been evaluated, but with no conclusive evidences, particularly due to the strong possibilities of common exposition to a rodent source (Toro et al. 1998).

Moreover, up to now, Andes virus is the only hantavirus undoubtedly linked to interhuman transmission, and no such evidence has been registered for the other New World hantaviruses. In Argentina, different evaluations did not find new evidence of interhuman transmission up to the end of 1999. An analysis of 197 contacts of 15 cases that occurred in 1999, associated with the genotypes Lechiguanas and Hu36964, did not find evidence of interhuman transmission (Colavecchia et al. 1999). On the other hand, a study was performed with 539 contacts, including 78 healthcare workers, of 56 HPS cases coming from the three endemic areas of Argentina between 1994 and 1999. In only 2 out of 380 contacts from the central area was an enzyme-linked immunosorbent assay (ELISA), IgM-positive reaction found. Both persons developed a mild febrile syndrome, not fulfilling HPS case definition. No evidence of interhuman transmission was found, and the most plausible explanation in both circumstances was high rodent infestations of domestic and peridomestic settings (Pini et al. 1999).

A review of the US HPS case registry at CDC, comprising 160 confirmed patients, identified five clusters of cases. In all of them, common exposure to a rodent-infested area was the most likely possibility. On this review, only one patient had a medically related occupation, but again contact with infected rodents was the most probable source of infection (WELLS et al. 1997b).

6.2 Infection in Children

For the United States, where most of the infections are caused by SNV, 4.5–6.9% of all reported infections have been in children less than 17 years old (BRYAN et al. 1997; PETERS 1998). The youngest infected person was 4 years old; the disease was mild, not fulfilling HPS case definition. In children, the clinical course of the disease and case-fatality rates have been similar to those reported in US adults.

Both Argentina and Chile have reported a higher percentage of children affected by hantavirus infections. In Argentina, cases in children came from the three ecologically different and geographically distant regions considered endemic for HPS. The official register up until 1998 included 12 cases younger than 14 years of age, and 18 younger than 17 years, representing 8% and 12% of the total, respectively. In 1999, of 55 confirmed cases, five (9%) were under the age of 14 and eight under age 17 (14.6%) (source: Ministerio de Salud y Acción Social). The youngest infected person was a 1-year-old who had a mild disease, not fulfilling HPS definition. Again, for children the clinical course of the illness and case-fatality rates were similar to those reported for adults (PINI et al. 1998).

Chile has also reported a relatively large proportion of children developing the disease, with a percentage of children under age 17 reported to be as high as 15.2 in the 1996 outbreak. The percentage of children in endemic situation seems to be lower (TORO et al. 1998).

6.3 Spectrum of Illnesses

For SNV, it is considered that most infections result in HPS. During the 1993 epidemic, and among the household contacts of HPS cases, a search for mild disease was unsuccessful. Seroprevalence studies also showed low percentages (less than 3%) among these groups and other normal persons. However, a recent report described four atypical mild cases that were identified through the US National Surveillance System, although they did not fulfill HPS case definition. It is difficult to accurately ascertain a case-fatality ratio because it has been decreasing from more than 50% in 1993 to 40%, and with a register in 1997, to 20% (PETERS 1998; KITSUTANI et al. 1999).

In Paraguay and northern Argentina, high human seroprevalence ranging from 7 to 21% among asymptomatic groups and community residents has been reported (WILLIAMS et al. 1997; S. Levis, unpublished data). Moreover, a study done on Indians living in western Paraguay and in the northeastern part of Salta province in

Argentina (Gran Chaco territory) found antibody prevalence as high as 40% (FERRER et al. 1998). This was considered a suggestion of a milder illness, and a much higher rate of subclinical infection with less pathogenic serologically cross-reactive hantaviruses.

In central Argentina it has been registered that 13% of cases are mild (BRIGGILER et al. 1998). Similarly, percentages of mild disease for Andes virus ranging from 11 to 13% have been reported (TORO et al. 1998; LÁZARO et al. 2000).

6.4 Seroprevalence of Antibodies in Humans

Studies of seroprevalence of antibodies in the United States show low seroprevalence rates, estimated to be overall ~0.3% (PETERS 1998). This pattern seems to be similar to the one found in Brazil, where extensive serosurveys have shown a relatively low prevalence of antibodies in several areas (IVERSSON et al. 1994). In Chile, IgG ELISA antibodies reactive with SNV were found in 2–13% of local residents (TORO et al. 1998; VALDERRAMA et al. 1998). In Argentina, low seroprevalence rates ranging from 0.1 to 1.5% were found in central and southern areas (LEVIS et al. 1995; WELLS et al. 1998). However, higher prevalence (up to 20%) of IgG ELISA antibodies reactive with SNV was registered in the northern subtropical area of the country.

In Paraguay, high human seroprevalence of antibodies (ranging from 7 to 21%) among asymptomatic groups and community residents was found. In another study among Indians living in the Gran Chaco region, comprising western Paraguay and Salta province of Argentina, up to 40% of the Paraguayan Indians and 18% of Argentine aborigines had antibodies detected by an immublot assay (FERRER et al. 1998).

6.5 Clinical Findings Associated with Other New World Hantaviruses

Bayou. In three of the Bayou-associated illnesses, renal compromise not requiring hemodialysis was registered, being severe in one patient who died, and mild in the other two. Cases came from Louisiana and Texas, USA (PETERS 1998).

Black Creek Canal. Only one human infection has been identified, and registered differences with classical SNV infection include renal failure not requiring hemodialysis and modest elevation of serum creatinekinase. Case originated from Florida, USA (PETERS 1998).

Juquitiba. The first three documented hantavirus infections in Brazil led to the identification of this new hantavirus. One of the cases was described as having a mild disease, with prominent vertigo (PETERS 1998).

Laguna Negra. Registered case-fatality rates for confirmed cases was 9%, but it is supposed to be higher considering the registered rate for confirmed and presumptive cases (29.4%). In the same area in which this virus was circulating, high human seroprevalence was found, as mentioned above, ranging from 7 to 21% (PETERS 1998). This was considered a suggestion of a milder illnesses and a much higher rate of subclinical infection than with the other viruses.

Lechiguanas. This is one of the two hantavirus viral genotypes identified causing human illnesses in central Argentina. Human infections in this area were found: (1) under the surveillance of Argentine Hemorrhagic Fever program, (2) testing for hantavirus infection all suspected leptospirosis cases, and (3) using HPS-case definition. In a study of 30 cases confirmed as hantavirus infections by serology or RT-PCR (reverse transcription-polymerase chain reaction), and hospitalized at INEVH during the years 1987–1999, it was possible to define a spectrum of clinical illnesses associated with Lechiguanas infections. In this analysis, patients were considered to have renal compromise if they had presented oliguria (less than 500ml/day), or creatinine levels of more than 20mg/l. Four patients (13%) had a mild disease (lasting approximately 1 week), characterized by fever, myalgia, headache, nausea, and vomiting, without significant findings on physical examination. One surviving patient (3%) had fever, nausea, vomiting, and renal compromise not requiring hemodialysis. Another single surviving patient (3%) had a febrile syndrome followed by an encephalopathy, characterized by confusion, excitation, and generalized convulsions. He had been considered under clinical basis as having a severe neurological form of Argentine hemorrhagic fever. These cases, which represent 20% (6/30) of this series, would be undetected by normal HPS surveillance. The remaining 24 patients (80%), fulfilled classical HPS, but with 50% (12/24) having renal compromise, as defined above. None of them required hemodialysis. The total case-fatality rate was 20% (6/30). For cases comprised under HPS case-definition, the case-fatality rate was 25% (6/34), being 33% (4/12) for HPS patients with renal compromise and 17% (2/12) for patients without renal compromise. Lechiguanas genotype was identified in 11 patients distributed among all clinical forms (Table 2) (Briggiler et al. 1998).

In the clinical laboratory, all cases had thrombocytopenia, with 54% presenting minimum values below 50,000/mm^3, and 96% under 100,000/mm^3. An increased hematocrit (female >46%; male >52%) was seen in only 28% of patients. White cell counts lower than 4,000/mm^3 were seen in only 14% of cases. The absence of leukopenia was useful in the differential diagnosis with Argentine hemorrhagic fever in the prodromal phase.

Hu39694. This is the other genotype responsible for human infections in central Argentina. The original case from which the genotype was identified has an HPS with renal compromise, not requiring hemodialysis, and survived.

Orán. This is a genotype found to be associated with human illnesses in the north of Argentina. In different reports of cases coming from this area, renal compromise was found in approximately 50% of cases. The case-fatality rate in the area has ranged between 18 and 40%, with a mean of 26% (Cortés et al. 1994).

Andes. This is the genotype responsible for hantavirus infections in southern Argentina and Chile. In a series of 19 Andes-associated cases coming from southern Argentina, the main clinical characteristics and radiological findings were similar to those reported above in the original description of HPS. However, in this group of patients, conjunctival congestion was referred in 47% (9/19), facial flushing in 7/19 (36.8%), pharyngeal congestion in 5/19 (26.3%), and petechiae in 3/19 (15.8%). On the other hand, urine analyses were abnormal in 11/11 cases and were characterized by presence of proteins, red blood cells, and granular

Table 2. Clinical spectrum of hantavirus infections in central Argentina

	Clinical forms					
	MFS[a]	FSRC[b]	FSE[c]	HPS[d]	HPS-RC[e]	Total
Number/percentage	4 (13%)	1 (3%)	1 (3%)	12 (40%)	12 (40%)	30 (100%)
Case-fatality rate	0%	0%	0%	17% (2/12)	33% (4/12)	20% (6/30)
Genotype (number of cases)	LECH[f] (2)	LECH (1)	LECH (1)	LECH (1)	LECH (6) Hu39694 (1)	LECH (11) Hu39694 (1)

[a] MFS: Mild febrile syndrome;
[b] FSRC: Febrile syndrome with renal compromise;
[c] FSE: Febrile syndrome with CNS compromise;
[d] HPS: HPS without renal compromise;
[e] HPS-RC: HPS with renal compromise;
[f] LECH: Lechiguanas.

cast. Mean serum creatinine levels were 27.5mg/l (range 8.1–110mg/l), with serum creatinine levels being higher than 20mg/l in 8/15 patients (53%). Two patients required hemodialysis. Hemoconcentration, expressed by an increased hematocrit, was found in only 25% of cases. One patient of this series had hemorrhages characterized by hematemesis, petechiae, and generalized hematomas. A second patient presented with a jugular hematoma. This patient would not fulfill original CDC case-definition because he was not previously healthy. He was an HIV-positive patient (CDC category B3) treated with zidovudine and didadosine (Lázaro et al. 2000).

7 Differential Diagnosis

During the prodromal phase it is difficult to differentiate HPS from other acute febrile conditions (Moolenaar et al. 1995; Koster and Levy 1998). When abdominal or back pain is severe, other possible diagnoses such as pyelonephritis, appendicitis, abdominal abscess, or gynecological infection should be considered. Once the cardiopulmonary phase begins, differential diagnosis includes bilateral pneumonia with sepsis; adult respiratory distress syndrome complicating systemic infections, including mycoplasma and influenza A, trauma, and other life-threatening conditions; and, sepsis syndrome complicated by either disseminated intravascular coagulation or alcohol toxicity. In North America, other zoonotic diseases prevalent in rural settings should be considered, including plague, tularemia, Rocky Mountain spotted fever or murine typhus, granulocytic or monocytic ehrlichiosis, leptospirosis, relapsing fever due to *Borrelia hermsi*, and acute parvovirus infection. In South America, other possible diagnoses include dengue, dengue hemorrhagic fever, and arenavirus hemorrhagic fevers (Junin, Machupo, Sabia, and Guanarito viruses). Figure 2 shows the algorithm proposed for early case recognition. It might be particularly useful in all areas with previously known or suspected cases of HPS (Pan American Health Organization 1999).

8 Etiological Diagnosis

The most feasible way of performing the etiological diagnosis of a hantavirus infection in humans is the serological method, because almost all confirmed cases have detectable IgM in the first or second sample after hospitalization.

ELISA IgM captured assay, using either SNV, Laguna Negra, or Andes antigens, are used in all countries that have detected cases. ELISAs are also performed to detect seroconversion by IgG in acute and convalescent-paired samples. Other serological tests in use include immunofluorescence and particle agglutination assay.

RT-PCR, although not widely available, is also useful as a diagnostic tool. The demonstration of hantavirus antigen in tissues by immunohistochemistry is a valuable method for laboratory diagnosis, particularly in fatal cases in the absence of serum or clot samples.

9 Pathology and Pathophysiology

In SNV infections, the most consistent histopathological features were observed in the lung and spleen. In the lungs, the basic histopathological finding was that of an interstitial pneumonitis with variable amounts of mononuclear cell infiltrates, congestion, and both interstitial and intra-alveolar edema. In typical HPS cases, hyaline membranes are focal, composed mainly of fibrin without cellular debris. Reactive type II pneumocytes and neutrophils are not prominent histological features. Large immunoblasts in the red pulp and in the periarteriolar lymphocytic sheath of spleen were consistently found. Endothelial cells were shown to have a remarkably high viral load, with hantaviral antigens consistently seen in capillary endothelial cells in various tissues throughout the body. The extent of pulmonary endothelial involvement was remarkable and far greater than previously seen with other hantaviral syndromes. Little evidence of endothelial cytopathic effect was found by light microscopy. In a series of autopsies, the microvasculature of kidney

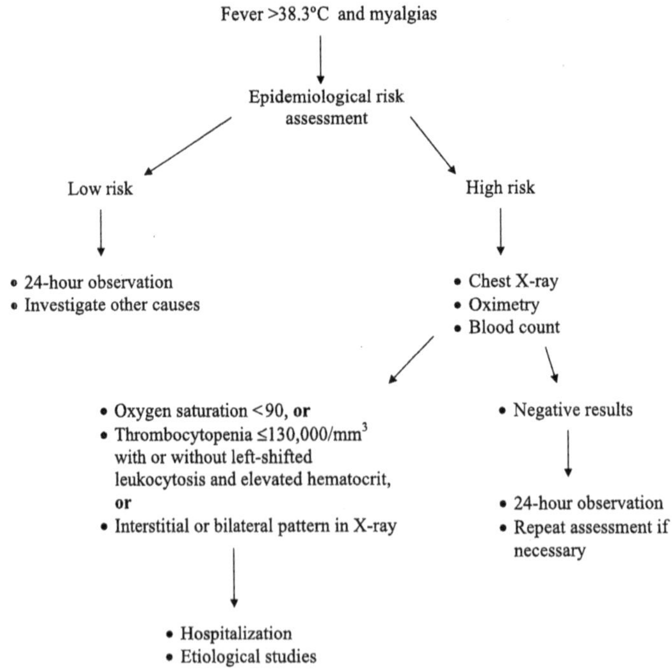

Fig. 2. Proposed algorithm for early diagnosis of HPS, which may be especially useful for areas where the disease has been confirmed or suspected

showed appreciable amounts of SNV antigens in the medulla and glomeruli; but insignificant amounts were seen in tubular epithelial cells. Despite little histological evidence of tissue damage, extensive deposits of hantaviral proteins were seen in the microvasculature of cardiac tissues (Zaki et al. 1995).

Reported pathological findings in a series of 12 incomplete autopsies of Argentine cases showed similar alterations to those previously described, and other lesions unreported for SNV infections. These include hepatic lesions of varying degree, from mild hepatocitic vacuolar changes to centrolobulillar necrosis; renal lesions ranging from tubular dilatation to wide necrosis of medullary tubules; and cardiac lesions, consisting of a mononuclear miocarditis. Immunohistochemical studies showed a similar distribution of viral antigens to those described for SNV patients (Lázaro et al. 2000; N. Pini and E. Neffen, unpublished observations). Also, in a report of Argentine cases associated with Andes virus, more dense infiltration of the alveolar interstice, more prominent hyaline membrane, with several displaying type T II pneumocytic activation and less airspace fibrin exudation than seen in North American cases were also described (Mansilla et al. 1997). In a description of the histopathological findings in Chilean cases from 1997 outbreak, a fine granular immunostaining in the hepatic cells unobserved in North American HPS cases was informed (Toro et al. 1998).

The lack of histological lesions to explain disordered organ function and death had led to the idea that hantaviruses induce altered endothelial function without overt cell death. In HPS, pulmonary edema plays an essential role in the fatal outcome of the disease and functional derangement of vascular endothelium appears to be central to the pathogenesis. The basis behind this alteration is thought to be mainly the consequence of the immune response to the viruses. Lymphoid infiltrates in the lungs has been described as a mixture of T lymphocytes, particularly those expressing CD8, and macrophages. Available data support the hypothesis that cellular immunity mediated by activated T cells may result in lung injury in HPS (Simpson and Levy 1999).

10 Treatment

There is no specific effective therapy for New World hantavirus infections, and treatment measures are mainly supportive. Ribavirin has been proved to have an in vitro antiviral effect to the members of the *Bunyaviridae* family, including SNV, and has been effective in the treatment of HFRS (Huggins et al. 1991). The efficacy in the treatment of HPS is unclear (Chapman et al. 1994). A placebo-controlled clinical trial is currently underway in the United States and Chile. One of the problems limiting the therapeutic possibilities of Ribavirin in HPS is the difficulty in initiating therapy early in the prodromal phase. In Argentina, a protocol for evaluating Ribavirin early in the course of the disease has been approved. The study was designed considering that in Argentina there were cases presenting in clusters, and the drug is to be evaluated in high-risk contacts of confirmed HPS cases that develop a febrile syndrome of unknown etiology during the 4 weeks after the contact.

In the absence of a proven effective therapy, the treatment of HPS is based on careful fluid management, hemodynamic monitoring, and ventilatory support. A broad spectrum antibiotic therapy (such as intravenous ceftriaxone or ampicillin-sulbactam) should be complemented due to the variety of differential diagnosis. This therapy should also include doxycycline until definite diagnosis is made. In the United States ECMO (extracorporeal membrane oxygenation) and inhaled NO have been used as a therapeutic salvage options, but the available data are still insufficient to analyze their efficacy (Hallin et al. 1996; Simpson and Levy 1999). Steroids have also been used in some patients with HPS, particularly in South America, with the hypothesis that the modification of the immune response may be of benefit. Their role in the treatment of this syndrome is unclear, and unless a careful evaluation adequately designed is made, no conclusions would be obtained.

11 Concluding Remarks

Over the past 10 years we have been confronted with an unusual explosion of knowledge in the area of hantaviruses. The original description of HPS in the United States in 1993 prompted the recognition of outbreaks as well as of isolated cases in other countries of the Americas, and many SNV-related hantaviruses were recognized, some of them not yet associated with human diseases.

Most of the human hantavirus infections in the New World have been identified using the CDC case definition that considers a "probable case of HPS to be a febrile illness (fever greater than 38.3°C), in a previously healthy person, characterized by unexplained acute respiratory distress syndrome or bilateral interstitial pulmonary infiltrates, with respiratory compromise requiring supplemental oxygen; or an unexplained illness resulting in death in conjunction with an autopsy examination demonstrating non-cardiogenic pulmonary edema without an identifiable specific cause of death". Confirmed case-patients must have laboratory evidence of hantavirus infection, shown by the presence of hantavirus-specific IgM antibodies, seroconversion by IgG or a positive RT-PCR result for hantavirus RNA, or a positive immunohistochemical result for hantavirus antigen (CDC 1997).

It has been suggested that the clinical presentation of some cases in South America were different from the classical description of HPS due to SNV. Also, some important epidemiological dissimilarities were also recorded. In this respect, the most outstanding one was the conclusive presence of interhuman transmission in the 1996 outbreak in El Bolsón, Argentina, due to Andes virus. To date, this has been the only situation in which there is convincing data, although person-to-person transmission was also suspected in other clusters of human illnesses associated with Andes virus. No other evidence of interhuman transmission has been found for the other hantaviruses so far. It cannot be predicted if this only represents an isolated, rare event merely linked to Andes virus, or if this could be a more generalized phenomena. Available evidence tends to favor the former hypothesis, but only careful prospective follow-up of cases and integrated studies (clinical, epidemiological, virological, molecular, and ecological) might eventually help to solve this point. In any event, this single episode calls for a strict adherence to biosafety measures when caring these patients, including additional respiratory precautions whenever indicated.

It seems also from the scarce available data that the spectrum of clinical illnesses may differ between regions, with areas sustaining higher percentages of milder infections. Case-fatality rates have been shown to fluctuate between countries and regions, and between years in the same region. For this reason, an overall estimation of the case-fatality rate for all New World hantaviruses should range between 20% and 50%. If these differences reflect dissimilarities in surveillance, or in the virulence or pathogenicity among different genotypes, or they are in relation to host factors, remains to be further investigated. The wide reported range in case-fatality rates would announce difficulties in the design, evaluation, and comparison of treatment protocols and would make it difficult to ascertain efficacy unless prospective randomized placebo-control studies are performed.

Additionally, a wide range of antibody positivity among humans has been reported, being as low as 1% in different areas of the United States, Brazil and Argentina, and as high of 40% in aborigines of Gran Chaco region of Paraguay. Explanations for these high seroprevalence rates may include the possibility of infections with less pathogenic serologically cross-reactive hantavirus, the presence of milder illnesses, or a much higher rate of subclinical infection than with the other viruses. In the case of the Gran Chaco Indian populations, it has been suggested that these populations might be frequently exposed to New World hantaviruses. The higher percentages of children reported in both Chile and Argentina might also relate to a population more frequently exposed to rodents.

Another reported difference between clinical manifestations of SNV-associated illness is the degree of renal compromise. In this respect, the cases associated with Bayou, Black Creek Canal, Andes, Lechiguanas, Hu39694, and Orán genotypes seem to exhibit more frequent renal compromise. This might reflect particular characteristics of some genotypes, but further studies will be needed to definitively address this point.

The integration of the surveillance of hantavirus-associated illnesses in the context of viral hemorrhagic fever syndrome has allowed the detection not only of milder forms of disease, but of unusual clinical forms, such as those with renal compromise without respiratory involvement, or a disease that resembles primary central nervous system. This is an expected finding considering the spectrum of clinical manifestations described in the large series of cases affected by HFRS.

In clinical laboratory studies, thrombocytopenia has been a common denominator in almost all cases from the Americas. This might not be the case for an increased hematocrit, which is considered a good marker for SNV-associated HPS, and was less frequently found in cases associated with Lechiguanas and Andes genotypes. Reasons for these differences remain to be ascertained.

Reported pathological findings of South American cases showed similar alterations to those described for SNV infections, and other hepatic, renal and cardiac lesions not described for the North American cases. It is still unknown if these differences would be consistently seen. They may be explained through the effect of myocardial depression, the hypoxia, or they might correlate to differences among hantavirus genotypes. Although the studies with other New World hantaviruses are insufficient, and despite the reported clinical and histopathological differences, preliminary studies reinforce the original idea of a similar pathogenesis of the human infection for all viruses coming from sigmodontine reservoirs, with pulmonary edema and suffocation being a key issue in most cases.

It is difficult to conclusively address the whole span of any clinical entity when only around 600 cases, distributed on seven different countries, in different settings, and under different observational conditions, have been registered. Any single event today constitutes an important percentage under a low denominator, but in the long term, it could eventually become an unusual or atypical presentation of the disease. On the other hand, it might end up being the original description of a more generalized phenomenon. With a better identification of the disease in the rest of the Americas, where undoubtedly there are cases occurring without recognition, we hope that some of today's questions will have the opportunity to be unfolded.

References

Briggiler A, Levis S, Pini N, Riera L, Enria D (1998) Presentación clínica de las infecciones por hantavirus en el área central de Argentina. Medicina (Buenos Aires) 52(5/2):673

Bryan RT, Doyle TJ, Moolenaar RL, Pflieger AK, Khan AS, Ksiazek TG, Peters CJ (1997) Hantavirus pulmonary syndrome. Semin Pediatr Infect Dis 8:44–49

Centers for Disease Control and Prevention (1997) Case definitions for infectious conditions under public health surveillance. Morbid Mortal Weekly Rep 46(RR-10):16

Chapman LE, Mertz G, Khan AS, Hart DC, Peters CJ, Koster E, Ksiazek TG, Rollin PE, Wilson L, Baum KF, Pavia AT, Christerson JC, Allen S, Rubin PJ, Goad D, Ribavirin Study Group (1994) Open label intravenous ribavirin for hantavirus pulmonary syndrome. Abstracts of the 34th Interscience Conference on Antimicrobial Agents and Chemotherapy, abstr. H-111. American Society for Microbiology, Washington DC; 1994

Colavecchia S, Martínez P, Miguel S, Edelstein A, González Della Valle M, Trincheri A, Marone J, Adala R, Verzeri L, Segura E, Rabinovich D, Padula P (1999) Distribución y variabilidad genética de los virus causantes de Síndrome Pulmonar por Hantaviurus (SPH) en la provincia de Buenos Aires. VI Congreso Argentino de Virología, Buenos Aires, 29–31 de Agosto de 1999. Abstract 153, p 61

Cortés J, Cacace ML, Seijo A, Parisi MN, Ayala LT (1994) Distress respiratorio del adulto en Orán, Salta. Presented in the 11er Simposio Internacional de Infectología Pediátrica, Córdoba, Argentina, 7–11 May 1994

Duchin JS, Koster FT, Peters CJ, Simpson GL, Tempest B, Zaki SR, Ksiazek TG, Rollin PE, Nichol S, Umland ET, Moolenaar RL, Reef SE, Nolte KR, Gallaher MM, Butler JC, Breiman RF, Hantavirus Study Group (1994) Hantavirus pulmonary syndrome: a clinical description of 17 patients with a newly recognized disease. N Engl J Med 330:949–955

Enria DA (1998) Emergencia de los Hantavirus en las Américas y en la Argentina. Medicina (Buenos Aires) 58(1):13–18

Enria DA, Padula P, Segura EL, Pini N, Edelstein A, Riva Posse C, Weissenbacher MC (1996) Hantavirus pulmonary syndrome in Argentina: possibility of person-to-person transmission. Medicina (Buenos Aires) 58:709–711

Ferrer JF, Jonsson C, Esteban N, Galligan D, Bosombrio MA, Peralta-Ramos M, Bharadwaj M, Torres-Martínez N, Callahan J, Segovia A, Hjelle B (1998) Epidemiological features of SNV-related hantavirus infection in high prevalence Gran Chaco Indian populations. Abstracts of the fourth International Conference on HFRS and Hantaviruses, p. 144

Hallin GW, Simpson SQ, Crowell RE, James DS, Koster FT, Mertz GJ, Levy H (1996) Cardiopulmonary manifestations of hantavirus pulmonary syndrome. Crit Care Med 24:252–258

Huggins JW, Hsiang CM, Cosgriff TM, Guang MY, Smith JI, Wu ZO, Le Duc JW, Zheng ZM, Meegan JM, Wang QN, Oland DD, Gui WE, Gibbs PH, Yuan GH, Zhang TM (1991) Prospective, double-blind, concurrent, placebo-controlled clinical trial of intravenous ribavirin therapy of hemorrhagic fever with renal syndrome. J Infect Dis 164:1119–1127

Iversson LB, Travassos da Rosa APA, Rosa MDB, Lomar AV, Sasaki MDGM, LeDuc JW (1994) Human infection by Hantavirus in southern and southeastern Brazil. Rev Ass Med Brasil 40(2):85–92

Johnson AM, Bowen MD, Ksiazek TG, Williams RJ, Bryan RT, Mills JN, Peters CJ, Nichol ST (1997) Laguna Negra virus associated with HPS in western Paraguay and Bolivia. Virology 238:115–127

Ketai LH, Williamson MR, Telepak RJ, Levy H, Koster FT, Nolte KB, Allen SE (1994) Hantavirus pulmonary syndrome: radiographic findings in 16 patients. Radiology 191(3):665–668

Khan AS, Spiropoulou CS, Morzunov S, Zaki SR, Kohn MA, Nawas SR, Mc Farland L, Nichol ST (1995) A fatal illness associated with a new hantavirus in Louisiana. J Med Virol 46:281–286

Khan AS, Gaviria M, Rollin PE, Hlady WG, Ksiazek TG, Armstrong LR, Greenman R, Ravkov E, Kolber M, Anapol H, Sfakianaki ED, Nichol ST, Peters CJ, Khabbaz RF (1996a) Hantavirus pulmonary syndrome in Florida: association with the newly identified Black Creek Canal virus. Am J Med 100:46–48

Khan AS, Khabbaz RF, Armstrong LR, Holman RC, Bauer SP, Graber J, Strine T, Miller G, Reef S, Tappero J, Rollin PE, Nichol ST, Zaki SR, Bryan RT, Chapman LE, Peters CJ, Ksiazek TG (1996b) Hantavirus pulmonary syndrome: the first 100 US cases. J Infect Dis 173:1297–1303

Kitsutani PT, Denton RW, Fritz CL, Murray RA, Todd RL, Pape WJ, Frampton JW, Young JC, Khan AS, Peters CJ, Ksiazek TG (1999) Acute Sin Nombre Hantavirus infection without pulmonary syndrome, United States. Emerg Infect Dis 5(5):701–705

Koster F, Levy H (1998) Clinical manifestations and treatment of HPS. In: Ho Wang Lee, Charles Calisher, Connie Schmajhon (eds) Manual of hemorrhagic fever with renal syndrome and hantavirus pulmonary syndrome – WHO Collaborating Center for Virus Reference and Research (Hantaviruses). Asan Institute for life Sciences, Seoul

Ksiazek TG, Peters CJ, Rollin PE, Zaki S, Nichol S, Spiropoulou C, Morzunov S, Feldmann H, Sanchez A, Khan AS, Mahy BWJ, Wachsmuth K, Butler JC (1995) Identification of a new North American hantavirus that causes acute pulmonary insufficiency. Am J Trop Med Hyg 52:117–123

Lázaro M, Resa A, Barclay C, Calanni L, Lasala M, Padula P, Pini N, Enria D (2000) Síndrome pulmonar por hantavirus en el sur andino argentino. Accepted for publication – Medicina (Buenos Aires)

LeDuc J, Smith GA, Pinheiro FP, Vasconcelos P, Rosa E, Maiztegui JI (1985) Isolation of a Hantaan-related virus from Brazilian rats and serologic evidence of its widespread distribution in South America. Am J Trop Med Hyg 34:810–815

LeDuc J, Smith GA, Childs JE, Pinheiro FP, Maiztegui JI, Niklasson B, Antoniadis A, Robinson DM, Khin M, Shortridge KE, Wooster MT, Elwell MR, Ilberty PLT, Koech D, Rosa EST, Rosen L (1986) Global survey of antibody to Hantaan-related viruses among peridomestic rodents. Bull WHO 64 pp 139–144

Levis SC, Briggiler AM, Cacase M, Peters CJ, Ksiazek TG, Cortés J, Lázaro ME, Resa A, Rollin PE, Pinheiro FP, Enria DA (1995) Emergence of hantavirus pulmonary syndrome in Argentina. Am J Trop Med Hyg (Sup) 53(2):233

Levis S, Morzunov S, Rowe J, Enria D, Pini N, Calderon G, Sabattini M, St. Jeor S (1998) Genetic diversity and epidemiology of hantaviruses in Argentina. J Infect Dis 177:529–538

Lopez N, Padula P, Rossi C, Lazaro ME, Franze-Fernandez MT (1996) Genetic identification of a new hantavirus causing severe pulmonary syndrome in Argentina. Virology 220:223–226

Lopez N, Padula P, Rossi C, Miguel S, Edelstein A, Ramírez E, Franze-Fernández MT (1997) Genetic characterization and phylogeny of Andes virus and variants from Argentina and Chile. Virus Res 50:77–84

Maiztegui JI, Becker JL, LeDuc JW (1983) Actividad del virus de la fiebre hemorrágica de Corea o virus muroide en ratas del puerto de la ciudad de Buenos Aires. Medicina (Buenos Aires) 43:871

Mansilla C, Tapia M, Nolte K, Barclay C, Samengo L, Feddersen R, Hjelle B (1997) Pathological findings and distribution of viral antigens in tissues of patients with fatal Andes hantavirus infection. Am J Trop Med Hyg 57(3):45

Moolenaar RL, Dalton C, Lipman HB, Umland ET, Gallaher M, Duchin JS, Chapman L, Zaki SR, Ksiazek TG, Rollin PE, Nichol ST, Cheek JE, Butler JC, Breiman RF (1995) Clinical features that differentiate hantavirus pulmonary syndrome from three other acute respiratory illnesses. Clin Infect Dis 21:643–649

Nichol S, Spiropoulou C, Morzunov S, Rollin P, Ksiazek TG, Feldmann H, Sanchez A, Childs G, Zaki S, Peters CJ (1993) Genetic identification of a Hantavirus associated with an outbreak of acute respiratory illness. Science 262:914–917

Padula P, Edelstein A, Miguel SDL, López NM, Rossi CM, Rabinovich RD (1998) Hantavirus pulmonary syndrome (HPS) outbreak in Argentina: molecular evidence for person-to-person transmission of Andes virus. Virology 241:323–330

Pan American Health Organization (1999) Hantavirus in the Americas: guidelines for diagnosis, treatment, prevention, and control. Pan American Health Organization, Washington DC

Parisi MN, Enria DA, Pini NC, Sabattini MS (1996) Retrospective detection of clinical infections caused by hantavirus in Argentina. Medicina (Buenos Aires) 56(1):1–13

Peters CJ (1998) In: Sheld WM, Craig WA, Hughes JM (eds) Hantavirus pulmonary syndrome in the Americas. Emerging Infections 2. ASM Press, Washington DC, pp 17–64

Pini N, Resa A, Laime G, Lecot G, Ksiazek T, Levis S, Enria D (1998) Hantavirus infection in children in Argentina. Emerg Infect Dis 4:85–87

Pini N, Cacase M, Ubeid C, Trincheri A, Resa A, Peretti D, Bolpe J, Bogni L, Briggiler A, Ksiazek T, Levis S (1999) Pautas para el estudio serológico de contactos de Síndrome Pulmonar por Hantavirus (SPH). VI Congreso Argentina de Virología, Buenos Aires, 29–31 de agosto de 1999. Abstract 082 pp 62

Schmaljohn C, Hjelle B (1997) Hantaviruses: a global disease problem. Emerg Infect Dis 3:95–104

Simpson SQ, Levy H (1999) Hantaviruses. In: Raphael Polin, Peter F Wright, Marcel Dekker (eds) Viral Infections of the Respiratory tract. Inc pp 281–306

Toro J, Vega J, Khan A, Mills J, Padula P, Terry W, Yadón Z, Valderrama R, Ellis B, Pavletic C, Cerda R, Zaki S, Wun-Ju S, Meyer R, Tapia M, Mansilla C, Baro M, Vergara J, Concha M, Calderón G, Enria D, Peters CJ, Ksiazek T (1998) An outbreak of hantavirus pulmonary syndrome, Chile 1997. Emerg Infect Dis Vol 4(4):687–694, Oct–Dec 1998

Valderrama R, Vega J, Terry W (1998) Community serological survey of infection by hantavirus in the XI region, Aysen, Chile. Abstracts of the Fourth International Conference on HFRS and Hantaviruses p 155

Weissenbacher M, Merani MS, Hodara VL, et al. (1990) Hantavirus infection in laboratory and wild rodents in Argentina. Medicina (Buenos Aires) 50:43–46

Weissenbacher M, Cura E, Segura EL, et al. (1996) Serological evidence of human hantavirus infection in Argentina, Bolivia and Uruguay. Medicina (Buenos Aires) 56:17–22

Wells RM, Young J, Williams RJ, Armstrong LR, Busico K, Khan AS, Ksiazek TG, Rollin PE, Zaki SR, Nichol ST, Peters CJ (1997a) Hantavirus transmission in the United States. Emerg Infect Dis 3: 361–365

Wells RM, Sosa Stani S, Yadón Z, Enria D, Padula P, Pini N, Mills JN, Peters CJ, Segura EL, Hantavirus Pulmonary Syndrome Group for Patagonia (1997b) An unusual hantavirus outbreak in southern Argentina: person-to-person transmission? Emerg Infect Dis 3:171–174

Wells RM, Sosa Estani S, Yadón Z, Enria D, Padula P, Pini N, González Della Valle M, Mills J, Peters CJ (1998) Seroprevalence of antibodies to Hantavirus in health care workers and other residents of Southern Argentina. Clin Infect Dis, pp 27–28

Williams RJ, Bryan RT, Mills JN, Palma RE, Vera I, Velásquez F, Baez EM, Schmidt WE, Figueroa RE, Peters CJ, Zaki SR, Khan AS, Ksiazek TG (1997) An outbreak of hantavirus pulmonary syndrome in western Paraguay. Am J Trop Med Hyg 57:274–282

Zaki SR, Greer PW, Coffield LM, Goldsmith CS, Nolte KB, Foucar K, Feddersen RM, Zumwalt RE, Miller GL, Khan AS, Rollin PE, Ksiazek TG, Nichol ST, Mahi BWJ, Peters CJ (1995) Hantavirus Pulmonary Syndrome. Pathogenesis of an Emerging Infectious Disease. Am J Pathol 146:552–579

Clinical Characteristics of Hantavirus Infections on the Eurasian Continent

M. LINDERHOLM[1,2] and F. ELGH[3,4,5]

1 Hantavirus Disease on the Eurasian Continent

In 1933 and 1934 two Swedish physicians, working in northern Sweden independently of each other, observed a previously undescribed disease (MYHRMAN 1934; ZETTERHOLM 1934). It was characterized by a rapid onset of high fever, malaise, chills, headache, abdominal, back and often generalized pain, and a renal syndrome with proteinuria and oliguria followed by a diuretic phase. A spontaneous recovery followed all cases described. In 1945 Myhrman proposed nephropathia epidemica

[1] Department of Infectious Diseases
[2] Department of Medicine, Umeå University, 901 85 Umeå, Sweden
e-mail: mats.linderholm.us@vll.se
[3] Department of Virology, Umeå University, 901 85 Umeå, Sweden
e-mail: fredrik.elgh@ume.foa.se
[4] Department of Microbiology, Defence Research Establishment, 901 82 Umeå, Sweden
[5] Centre for Microbiological Preparedness, Swedish Institute for Infectious Disease Control, 171 82 Solna, Sweden

(NE) as the name for this disease. During the subsequent decades, a large number of cases of this disease were reported from Scandinavia and Finland (Myhrman 1945, 1951; Lähdevirta 1971).

During the Korean conflict (1951–1954) more than 3,000 United Nation soldiers acquired a severe disease characterized by fever, hemorrhagic manifestations and renal insufficiency (Gajdusek 1953, 1962; Smadel 1953; Sheedy et al. 1954). This disease, at that time largely unknown to western medicine, was recognized as the Korean hemorrhagic fever (KHF). Mortality rates were high, 5–10%, and mainly due to severe bleeding, shock, and renal failure.

However, evidence of a similar disease in the east was documented as early as 1913 in medical records from Vladivostok (Casals et al. 1970). Russian and Japanese literature, as reviewed by Gajdusek (1953) and Smadel (1953), also revealed the presence of a similar disease in the southeastern part of Russia and in Manchuria. Chinese authorities reported the existence of a similar disease in inner Mongolia (Gajdusek 1962). In the late 1950s, Soviet physicians reported diseases with similar features in European Russia. Identical disease patterns had been observed already in the 1930s under the name of Tula fever. A viral etiology was suggested by experiments from the 1940s in which Soviet and Japanese researchers provoked the disease in humans by parenteral injections of bacteria-free filtrates of body fluids (Gajdusek 1962). In 1953 Gajdusek postulated the relationship between Asian and European hemorrhagic fevers.

The emergence of KHF during the Korean War marked the starting point of intense studies in search of a causative agent. Circumstantial evidence pointed towards the role of rodents in the spread of these hemorrhagic fevers (Hortling 1946; Casals et al. 1970; Lähdevirta 1971; Nyström 1977). An agent that reacted strongly in indirect fluorescent antibody assay (IFA) to antibodies from patients that had acquired KHF was identified in lung sections from *Apodemus agrarius* (field mouse) rodents (Lee and Lee 1976, 1977; Lee et al. 1978), captured in rural endemic areas. The responsible virus, named Hantaan (HTN) after the original site of rodent capture at the Hantaan river in Korea, was isolated and could later be propagated in cell culture (Lee et al. 1978; French et al. 1981). This greatly simplified serological diagnosis of the disease. Soon thereafter, it was possible to show that sera of NE patients from Sweden and Finland reacted with an identical fluorescence pattern to the KHF antigen (Lee et al. 1979; Svedmyr et al. 1979). When similar NE antigen preparations derived from Finnish *Chlethrionomys glareolus* (bank vole) were available (Brummer-Korvenkontio et al. 1980) it was found that a high proportion of NE sera reacted with both agents (Svedmyr et al. 1980). Serological investigations performed in the former Soviet Union described evidence for the circulation of several distinct viruses responsible for similar diseases (Gavrilovskaya et al. 1981; Tkachenko et al. 1982). Furthermore, evidence of several serotypes in Europe was put forward the same year (Lee et al. 1982b). The NE-causing virus was isolated from lungs of *C. glareolus* adapted to growth in cell culture and named Puumala virus (PUU) after the original site of rodent capture (Brummer-Korvenkontio et al. 1982). Similar viruses have been isolated from *C. glareolus* and humans in Sweden (Juto et al. 1997; Niklasson and LeDuc 1984; Yanagihara et al. 1984).

A virus with immunological properties similar but distinct from HTN was isolated from *Rattus* in urban areas and has become known as the Seoul virus (SEO; LEE et al. 1982a). This virus has been shown to be pathogenic to man and represents a serotype of its own (CHU et al. 1994; LI et al. 1995). Antibodies directed towards this hantavirus have been detected throughout the world possibly due to the wide distribution of its *Rattus* host (LEDUC et al. 1986; CHILDS et al. 1988).

Infections caused by these viruses are major health problems on the Eurasian continent and cause disease in several hundred people annually. The name hemorrhagic fever with renal syndrome (HFRS) was recommended for the diseases within this group, and hantavirus became the accepted term for the causative viruses (WHO 1983; SCHMALJOHN et al. 1985). HFRS is now reported from all over the Eurasian continent. The major hantaviruses associated with HFRS and their main geographical distributions are summarized in Table 1. HTN and SEO virus have been found to cause a majority of the abundance of HFRS cases throughout the eastern part of the Eurasian continent (LEE et al. 1990). Human infection in the west has been reported to be caused by a number of hantaviruses. Lately not only PUU but also the recently discovered Dobrava virus (DOB) has conclusively (by reverse transcription-polymerase chain reaction, neutralization tests, or both) been shown to cause disease in Eurasia (AVŠIC-ZUPANC et al. 1992; HÖRLING et al. 1995; ANTONIADIS et al. 1996; LUNDKVIST et al. 1997b; PLYUSNIN et al. 1997; PAPA et al. 1998). PUU has been found to be the cause of HFRS all throughout Europe with high incidence in European Russia, Finland, and Northern Scandinavia. DOB has (in parallel with PUU) been shown to cause a large number of cases of severe disease on the Balkan Peninsula where this virus was originally isolated (AVŠIC-ZUPANC et al. 1992, 1999; ANTONIADIS et al. 1996; LUNDKVIST et al. 1997b). There are also sporadic cases reported from Central Europe of DOB-caused disease (LUNDKVIST et al. 1997a, 1998; MEISEL et al. 1998). Reports on HFRS caused by SEO- and HTN-like viruses also exist (AVŠIC-ZUPANC et al. 1994; CLEMENT et al. 1994b; ALEXEYEV et al. 1996). Whether these infections were caused by the postulated viruses or the antigenically and relatively closely related DOB awaits further investigation. The PUU-like Tula virus, found in central Europe, has also been associated with sporadic cases of HFRS (VAPALAHTI et al. 1996).

Before 1993, the common belief was that the American continent was free of human pathogenic hantaviruses except for cases of disease related to SEO harbored by rats in major ports. In 1993, an outbreak of a febrile pulmonary syndrome caused by hantavirus, and later named hantavirus pulmonary syndrome (HPS), changed this notion (NICHOL et al. 1993). At first this disease entity seemed

Table 1. Hantaviruses occurring in Europe and Asia presently known to cause disease in humans

Hantavirus serotype	Main rodent reservoir	Human disease	Main geographical distribution
Hantaan	*Apodemus agrarius*	HFRS (KHF)	East Asia
Seoul	*Rattus* spp.	HFRS (Urban)	Worldwide
Dobrava	*Apodemus flavicollis*	HFRS (KHF-like)	The Balkan region
Puumala	*Clethrionomys glareolus*	HFRS (NE)	Europe

unrelated to HFRS, but as comparisons between the two entities presented in this article show, a number of common features are evident.

2 Hemorrhagic Fever with Renal Syndrome: Clinical Aspects

As indicated by the name, the different forms of HFRS are febrile illnesses that generally include various degrees of hemostatic and renal disturbances. The clinical severity of HFRS ranges from asymptomatic infection to fulminant hemorrhagic shock and death. HTN and DOB virus are the causative agents of an illness with considerable mortality, whereas diseases caused by SEO and PUU viruses are less severe. However, the differences in clinical severity within illnesses caused by one virus serotype may be just as great as between the different serotypes. The cause of this variation is unknown. It may involve the virulence of individual viral strains, the infective dose, or host factors.

In typical cases of HFRS, five different phases are recognized (Sheedy et al. 1954). The disease usually begins abruptly with a flu-like, febrile phase. This is followed by a hypotensive phase, often associated with hemostatic disturbances ranging from petechiae and epistaxis to fatal intracranial hemorrhages. Renal impairment ensues, frequently accompanied by marked abdominal and/or back pain. First there is an oliguric phase, with declining urinary output and sometimes anuria, which is followed by a polyuric phase. Finally, there is a recovery phase that may be prolonged over weeks or months. In individual cases of HFRS, not all of these phases are always apparent and they may overlap considerably.

A hallmark of HFRS is vascular dysfunction, including vasodilatation and increased capillary permeability. Vascular dysfunction may explain many of the characteristic features of HFRS, such as hypotension, interstitial edema, and body cavity effusions and it may be of fundamental pathophysiological importance. The mechanisms by which hantaviruses initiate those vascular changes are, however, largely unknown.

2.1 Korean Hemorrhagic Fever

KHF, also known as epidemic hemorrhagic fever and hemorrhagic nephrosonephritis, is a severe form of HFRS. Each year between 50,000 and 100,000 patients are hospitalized in China alone (Lee et al. 1976, 1977; Chen et al. 1986), of which thousands are fatal infections. The more recently isolated Dobrava virus is the cause of a disease that does not seem to differ decisively from KHF (Antoniadis et al. 1987; Avšic-Zupanc et al. 1992, 1999). In the following text, illnesses caused by these two viruses will be referred to as KHF.

As previously indicated, five different phases of the disease are recognized (Sheedy et al. 1954): (1) febrile, (2) hypotensive, (3) oliguric, (4) polyuric, and

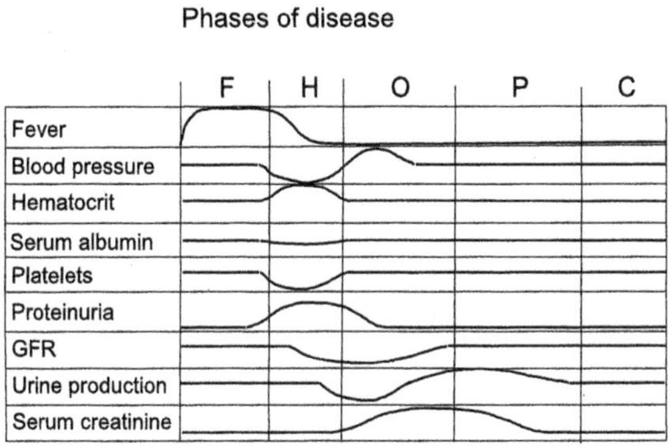

Fig. 1. HFRS, phases of disease. *F* Febrile, *H* Hypotensive, *O* Oliguric, *P* Polyuric, *C* Convalescence

(5) convalescent. Characteristic clinical and laboratory findings of different phases of the disease are summarized in Fig. 1. After an incubation period that may vary from 1 to 8 weeks, there is a sudden onset of high fever and chills (GILES et al. 1954; POWELL 1954; GAJDUSEK 1982; LEE 1989). High fever usually persists for more than 5 or 6 days and is accompanied by headache, myalgia, and nausea. Blurred vision is a typical complaint. Flushing of the face, an erythematosus rash, and conjunctival injections are characteristic clinical findings. Abdominal pain, back pain, or both ensure. At the end of the febrile period, the blood pressure declines and shock may occur. Increased capillary permeability is suggested by the coincidental appearance of hemoconcentration with normal to low serum protein concentrations and with edema, often located to the retroperitoneum. Thrombocytopenia is another characteristic finding and may contribute to the gradually appearing bleeding manifestations (e.g., petechiae of the skin and mucosa, epistaxis, and in severe cases, gastrointestinal or intracranial bleeding). In severe cases, prominent increase of white blood cell (WBC) numbers is observed, including activated lymphocytes and immature granulocytes. In the febrile phase, hematuria and proteinuria appear as the first signs of renal involvement, whereas glomerular filtration rate is essentially unaffected (FROEB et al. 1954). Generally, oliguria or anuria and rising concentrations of serum creatinine ensue, suggestive of a prominent renal involvement. Weight gain and rise of blood pressure, sometimes to hypertensive levels, follow. Polyuria heralds recovery. Diuresis, partly due to impaired tubular function, may be profound, and electrolyte imbalance may occur. The convalescence phase, finally, may be prolonged over weeks or months with a gradual recovery of renal function.

It should be remarked that the clinical expression of KHF varies considerably, and that only about one fifth of the patients develop severe hypotension and hemorrhages (LEE 1989).

2.1.1 Histopathological Findings

Lacking suitable animal models, much of the present knowledge on pathophysiology has been gained by autopsy and histopathological studies. The findings have varied according to the phase of disease investigated.

In autopsies of KHF patients, vascular abnormalities have been generally found, including vascular congestion, extravasation of red blood cells, and perivascular edema (LUKES 1954). In one study, degenerative changes of endothelial cells and of the basement membrane were demonstrated by use of electron microscopy (LI et al. 1984). In patients dying in the early, hypotensive phase of disease, retroperitoneal edema and effusions of the body cavities were observed (LUKES 1954). A typical triad of interstitial hemorrhage involved the right atrium, the anterior pituitary gland, and the renal medulla. Signs of vasculitis were virtually absent.

2.1.2 Renal Findings

On gross examination of fatal cases of KHF, the kidneys were enlarged with a pale cortex and a congested, hemorrhagic medulla. Also, by microscopic examination, the medullary region disclosed the most prominent changes, including congested peritubular capillaries, interstitial hemorrhages, mononuclear cell infiltrates, and interstitial edema. The tubular epithelium showed varying grades of degeneration, from flattening to necrosis. Some glomeruli showed capillary engorgement and focal thickening of the basement membrane (POWELL 1954).

2.1.3 Pulmonary Findings

In 4 of 11 patients dying within the first week after onset of KHF, autopsies showed pulmonary edema. This was also a striking finding in 16 of 28 patients dying at later (oliguric and polyuric) stages (LUKES 1954). Pleural effusions were described mainly in the hypotensive phase. At an early stage, edema was held to be associated with vascular leakage, whereas later in the acute phase of the disease, plasma volume overload due to renal failure and resorption of extra-vascular fluids may have contributed to edema formation (GILES et al. 1954; LUKES 1954). On microscopic examination, vascular dilatation and engorgement were universal findings and many of the alveoli were filled with eosin-stained exudate (POWELL 1954).

2.2 Urban HFRS

Like the rodent reservoir of SEO virus, *Rattus* species, the associated disease has a worldwide and mainly urban distribution (LEE et al. 1990). As compared with KHF, the illness caused by SEO virus has the characteristics of HFRS but is less severe with a mortality rate of approximately 1% (Table 2) (LEE 1989; KIM et al. 1995). Macroscopic hemorrhagic manifestations are rare. It is associated with more hepatic involvement than seen in KHF. Pulmonary effusions and dyspnea has been reported in SEO hantavirus-infected patients in North America (GLASS et al. 1994).

2.3 Nephropathia Epidemica

NE is most prevalent in the European and central parts of Russia, Finland, and Scandinavia. NE is also frequently reported in other parts of Europe.

Table 2. Some clinical and laboratory findings (%) in illnesses caused by Eurasian hantaviruses

	KHF[a]	Urban HFRS[b]	NE[c]
Fever	100	100	98–100
Systolic hypotension (< 90mmHg)	28–80	22	1–16
Pulmonary radiography pathology	13	nd	23–53
Internal hemorrhage	34	5	0–2
Thrombocytopenia (< 100×10^9/liter)	78	70	20–52
Hematuria	85	73	56–85
Proteinuria	100	100	94–100
Oliguria (< 500ml/d)	60–65	37	54–70
Dialysis	40	20	0–6

Data are approximated from references:
[a] KHF: Lee 1989, 1991; Powell 1954;
[b] Urban HFRS: Lee 1989;
[c] NE: Kanerva et al. 1996; Linderholm et al. 1992; Lähdevirta 1971; Mustonen et al. 1994a; Settergren et al. 1989.

Compared to KHF, NE is generally a milder form of HFRS (Table 2). Yet the clinical description of KHF is applicable. Like KHF, NE begins with a sudden onset of fever, chills, nausea, and headache (Lähdevirta 1971; Settergren et al. 1989). Frequently, abdominal and back pains ensue and may be severe. Hypotension is reported in up to 40% of the patients (Lähdevirta 1971) but is generally mild. Elevated WBC counts occur in NE, although not to the same magnitude as in KHF. Fatal cases of NE have been associated with marked leukocytosis and shock (Linderholm et al. 1991; Valtonen et al. 1995). In the early phase of disease, thrombocytopenia is noticed in a majority of the patients (Lähdevirta 1971). In a prospective study, bleeding manifestations and mild respiratory symptoms were reported in about one third of the patients (Settergren et al. 1989).

Beginning in the late febrile phase, urinalysis discloses proteinuria and hematuria in almost 100% of cases. Renal impairment follows, as indicated by oliguria and rising serum concentrations of creatinine and urea (Lähdevirta 1971; Settergren et al. 1989; Mustonen et al. 1994a). Severe hypertension, blood volume overload, or life-threatening electrolyte imbalance are rare events and dialysis is needed in a minority of patients only. It should be noted that NE does not always proceed with apparent renal affection (Alexeyev et al. 1993). In studies of serologically verified cases, no or minimally increased serum creatinine levels were reported in 3/74 (4%; Settergren et al. 1989) and 8/126 (6%; Mustonen et al. 1994a) of hospitalized patients, respectively. Thus, a normal concentration of serum creatinine does not exclude a diagnosis of NE.

2.3.1 Histopathological Findings

Only rare fatal cases of NE have been reported. In these instances, renal autopsy findings have been consistent with those of KHF (Linderholm et al. 1991; Forslund et al. 1992; Valtonen et al. 1995). There are, however, several studies of renal biopsy findings in NE. Here, changes within the medulla were most prominent and, as in KHF, included degeneration of tubular epithelial cells and interstitial edema with infiltration of red blood cells and leukocytes (Lähdevirta 1971). In a recent study of serologically verified cases of NE, the most common finding was an acute tubulointerstitial nephritis, whereas glomerular changes were reported in only 25% (Mustonen et al. 1994b).

3 Similarities Between HFRS and HPS

In 1993, a novel hantavirus infection was recognized in the United States. The clinical picture differed from HFRS and the syndrome was to be named HPS. The mortality rate was as high as 50%. The main difference from HFRS was the occurrence of a severe non-cardiogenic pulmonary edema early in the course of disease and less-severe renal complications.

It may seem surprising that hantaviruses causing HFRS and HPS are closely related, yet have so different clinical expressions, i.e., the predilection for renal and pulmonary involvement in HFRS and HPS, respectively. For example, Sin Nombre virus, associated with the highly lethal HPS, is serologically and genetically more closely related to Puumala virus, which causes a comparably mild disease, than to Hantaan virus, the etiological agent of the more severe form of HFRS.

However, the differences in clinical pictures between HFRS and HPS may not be as marked as it seems at first glance. The differences in organ manifestations may reflect only the major localization of injury imposed by similar underlying pathophysiological events. Both diseases are febrile illnesses with an acute onset. In both syndromes, a more generalized vascular involvement (capillary leak and vasodilatation) is apparent as demonstrated by hypotension, hemoconcentration, and vasodilatation. Other laboratory features in common include thrombocytopenia, proteinuria, and leukocytosis, with occurrence of activated lymphocytes in peripheral blood. The main clinical difference between the two conditions seems to be that the mainly afflicted vascular beds differ, in HFRS capillaries of the renal medulla, and in HPS pulmonary capillaries. Nevertheless, many of the basic clinical and laboratory aberrations are shared in the two conditions.

3.1 Renal Manifestations

Renal impairment is typical in HFRS. In addition, in HPS, and in particular HPS associated with Bayou and Black Creek Canal viruses, a modest to prominent renal involvement has been reported (Khan et al. 1995; Hjelle et al. 1996). Moreover, there are several reports of HFRS cases without apparent renal impairment (Settergren et al. 1989; Alexeyev et al. 1993; Mustonen et al. 1994a). Thus, the difference in renal dysfunction in HFRS and HPS may not be as clear-cut as generally perceived.

3.2 Pulmonary Manifestations

Non-cardiogenic pulmonary edema is a characteristic trait of HPS. The spectrum of the disease may, however, be wider. Recent reports suggests that HPS, associated with Sin Nombre virus, may proceed without apparent pulmonary symptoms (KITSUTANI et al. 1999; ZAVASKY et al. 1999).

In HFRS, respiratory tract symptoms are not usually included in the general description. Nonetheless, there is evidence of pulmonary involvement in the disease. In the original reports of HFRS in 1954 among 828 patients, pulmonary complications were recorded in 6% (GILES et al. 1954) and fatal pulmonary edema in 2% (LUKES 1954). In these reports, pulmonary complications were, in fact, one of the main causes of death. Among 181 patients subjected to radiological examination, pneumonic infiltrates were found in 3% and congestive changes in 10% (POWELL 1954). These data all involve the Hantaan virus-caused form of HFRS. In spite of this, there have been few studies focusing on the pulmonary aspects of HFRS, possibly because other, more prominent symptoms, e.g., vascular, hemostatic, and renal disturbances, dominate the clinical picture.

More recent studies, however, have shown that pulmonary manifestations can be demonstrated even in mild HFRS. In NE, caused by Puumala virus, pulmonary edema has been reported only rarely (CLEMENT et al. 1994a; KANERVA et al. 1996). Among a variety of general symptoms in the acute phase of NE, cough has been recorded in 6–32% of the cases. (LÄHDEVIRTA 1971; SETTERGREN et al. 1989; MUSTONEN et al. 1994a). In three Scandinavian studies, radiological pulmonary changes were noted in 27/97 (28%), 10/19 (53%), and 13/55 (23%), of the cases (LÄHDEVIRTA 1971; LINDERHOLM et al. 1992; KANERVA et al. 1996). The radiographic changes included interstitial infiltrates in the perihilar tract or the peripheral and/or lower parts of the lungs, as well as pleural effusions. There was an association of radiological findings with the severity of the clinical condition (LÄHDEVIRTA 1971) as well as with parameters of inflammation (LINDERHOLM et al. 1992; KANERVA et al. 1996).

In bronchoalveolar lavage (BAL) fluid of five patients in the acute phase of NE, increased numbers of macrophages, activated macrophages, CD8 T cells, and NK cells were found (LINDERHOLM et al. 1993). Numbers of CD4 T cells and B cells were normal. Similar to HPS (ZAKI et al. 1995), and in contrast to the adult respiratory distress syndrome (ARDS), only a modest accumulation of neutrophils occurs in the lower respiratory tract in the acute phase of NE (LINDERHOLM et al. 1993). The fibronectin concentration in BAL fluid was higher in patients with NE than in a control group. Altogether, a local host response in the lower respiratory tract was found in NE compatible with what might be expected to occur in viral disease. Moreover, the pulmonary changes in NE may be similar by nature to those seen in HPS.

The cardiac and pulmonary functions were investigated in 13 patients in the acute phase of NE (LINDERHOLM et al. 1997). Compared to reference values, the patients showed a decreased diffusion capacity for carbon monoxide and an increased pulmonary clearance of inhaled technetium-99m-labeled diethylenetriamine pentaacetic acid. In 4/11 patients, arterial hypoxemia was found. The cardiac function was normal. In sum, an alveolocapillary lesion best explained the pulmonary dysfunction.

Altogether, pulmonary changes seem to be far more common in HFRS than is usually recognized. It should be remarked that the illness of patients with HFRS referred to was relatively mild (Lähdevirta 1971; Linderholm et al. 1992, 1993, 1997; Kanerva et al. 1996). Thus, a pulmonary involvement seems to be a general trait of HFRS and not only an occasional finding in severe cases. Even though the occurrence of pulmonary manifestations in HFRS might seem to reconcile poorly with the relative sparseness of respiratory symptoms, it should be recalled that HFRS is associated with marked general symptoms for which an impaired pulmonary function may be partly responsible.

In summary, HFRS and HPS share many clinical characteristics. In both forms of illness, renal and pulmonary involvement may occur, even if the degree of severity varies. In spite of differences in clinical expression, basic pathogenetic mechanisms may be similar.

4 Prognosis

The overall mortality of KHF is 3–10% (Lee et al. 1990). In a report from the Korean War, the most common causes of death were irreversible shock and pulmonary complications (Lukes 1954). In nonfatal cases full recovery is considered the rule. Severe complications, such as neurological damage and panhypopituitarism may, however, develop because of intracranial hemorrhagic manifestations. Restoration of renal function is generally complete even though this may proceed over months. In SEO or SEO virus-like infections, a relationship between previous and chronic hypertensive end-stage renal disease has been reported in patients from the eastern United States (Glass et al. 1993). Of patients with end-stage renal disease because of hypertension, 6.5% were SEO virus seropositive.

In NE, there are only occasional reports of fatal outcome (Linderholm et al. 1991; Forslund et al. 1992; Valtonen et al. 1995), and overall mortality rate is $<0.5\%$ (Lee et al. 1990). As in KHF, long-term prognosis seems to be good (Lähdevirta 1971; Settergren et al. 1989), even if some patients may experience convalescence with fatigue lasting over months. Sequelae due to complications during the acute phase of NE are reported. This includes hemorrhagic complications (Alexeyev et al. 1994; Settergren et al. 1983; Zeier et al. 1992), Guillain-Barré syndrome (Esselink et al. 1994; Forslund et al. 1992) and other neurological complications (Alexeyev et al. 1995), and the development of pituitary insufficiency (Forslund et al. 1992; Settergren et al. 1992). Permanent renal injury is probably rare in NE (Lähdevirta 1971; Settergren et al. 1989). Tubular dysfunction, disclosed as impaired capacity to concentrate the urine, may persist for months. NE followed by hypertension has occasionally been reported (Kleinknecht and Rollin 1992). A similar investigation on Swedish patients could not reveal an overrepresentation of seropositivity for hantavirus (PUU virus) in end-stage renal disease (SETTERGREN et al. 1998).

After recovery from HFRS, long-lasting immunity follows and reinfection does not seem to occur (LEE 1982).

5 Diagnosis of HFRS Caused by Hantaviruses in Eurasia

The diagnosis of HFRS had been based on clinical and laboratory findings up until the point when the hantaviruses were found in lung tissue of rodents in various regions of the Eurasian continent (LÄHDEVIRTA 1971; LEE et al. 1976; NYSTRÖM 1977; BRUMMER-KORVENKONTIO et al. 1980; AVŠIC-ZUPANC et al. 1994). Clinical diagnosis is still the most important means of deciding if a patient has HFRS and the incentive for specific diagnostic testing. For specific serodiagnosis, hantavirus-infected lung sections were used as antigen in indirect IFA (LEE et al. 1976; BRUMMER-KORVENKONTIO et al. 1980). When hantavirus could be propagated in cell culture, this was used as antigen in IFA and ELISA (FRENCH et al. 1981; TKACHENKO et al. 1981). Lately, recombinantly produced hantavirus nucleocapsid proteins have been widely used in the enzyme-linked immunosorbent assay (ELISA) format for serological diagnosis of acute hantavirus infections and for epidemiological purposes (ROSSI et al. 1990; KALLIO-KOKKO et al. 1993; ZÖLLER et al. 1993; ELGH et al. 1996, 1997). Sera from hantavirus-infected patients react strongly to these proteins early in the course of disease (ELGH et al. 1997, 1998). A major drawback is the cross-reactivity seen between the different hantavirus serotypes. For serotypic differentiation, neutralization tests remain the most precise serological tool available (LEE et al. 1985; NIKLASSON et al. 1991; HÖRLING et al. 1992; CHU et al. 1994, 1995). An interesting approach for serotyping is under development, using non-cross-reactive parts of recombinantly produced nucleocapsid proteins of hantaviruses (MORII et al. 1998). Polymerase chain reaction specific for hantaviruses has been tried in HFRS but is not sensitive enough to be considered as a routine diagnostic tool (ARTHUR et al. 1992; HÖRLING et al. 1995; ANTONIADIS et al. 1996; PLYUSNIN et al. 1997). Most patients present specific IgM at onset of disease and this type of test in different formats is therefore the method of choice in HFRS diagnostics in endemic areas (ELGH et al. 1995, 1998). Moreover, specific IgA antibodies closely follow the kinetics of IgM and may provide additional diagnostic information (ELGH et al. 1998).

6 Treatment

The treatment of patients with HFRS relies on good supportive care. In severe cases of HFRS, intensive care facilities may have contributed to a reduction in mortality observed over the last decades.

Many drug therapies have been tried in HFRS. One strategy has been to use immunomodulatory drugs. For example, in the early phase of KHF, interferon-α (Gui et al. 1987) and steroids (Qian et al. 1990) were tried without evident effect. Cyclophosphamide, an immunosuppressive alkylating agent, administered early was shown to halt progression to hypotensive and oliguric stages (Dai et al. 1981). The only medication that has shown some promise is the antiviral agent ribavirin, which efficiently inhibits hantavirus growth in vitro (Kirsi et al. 1983). In a randomized study on KHF, ribavirin was reported to significantly reduce mortality when administered early after onset of disease (Huggins et al. 1991). The risk of hemorrhages and of entering the oliguric stage was also reduced in the ribavirin-treated group.

Still, proper supportive care remains fundamental in the treatment of HFRS. In the hypotensive and oliguric phases, fluid restriction is recommended and was in fact a measure reported to reduce the mortality among US soldiers afflicted with KHF (Steer 1955). In cases with severe hypotension, colloidal solutions and vasopressor agents are preferably used to maintain adequate blood pressure. Careful control of electrolyte balance is important. Antipyretic and analgesic agents are frequently needed. Drugs that may interfere with renal function, such as non-steroid anti-inflammatory drugs and renal toxic drugs, should be avoided. In severe forms of HFRS, dialysis may be required, particularly in the oliguric phase if hyperkalemia, severe hypertension, or hypervolemia occur. In NE, dialysis is rarely needed, even in cases with prominent elevation of serum creatinine concentration and profound oliguria (Lähdevirta 1971; Settergren et al. 1989).

7 Strategies for Future Treatment Trials

There are studies suggesting that host inflammatory responses are involved in the pathogenesis of HFRS and HPS. In particular, certain cytokines, such as tumor necrosis factor (TNF)-α, have been associated with hantavirus infection (Linderholm et al. 1996; Temonen et al. 1996; Mori et al. 1999). If cytokines can be shown to be among the determinants of the outcome of hantavirus disease, this should be a clue in the search for treatment of fulminant cases. Potential therapeutic interventions may be tried to down-regulate the inflammatory responses in severe cases of HFRS. Work to find inhibitors of TNF-α is in progress (Knight et al. 1993; Moreira et al. 1993; Evans et al. 1994). As a basis of future trials, it seems worthwhile to assay TNF-α and other cytokines in various forms of HFRS and in HPS.

Based on studies suggesting a favorable effect in KHF of ribavirin (Huggins et al. 1991), a combination of antiviral and immunomodulatory treatment would be particularly interesting. Inhibition of the inflammatory responses might possibly induce an exacerbation of hantavirus replication, an event that would be prevented by concomitant administration of antiviral drugs.

References

Alexeyev OA, Baranov BA (1993) Puumala virus infection without signs of renal involvement. Scand J Infect Dis 25:525–527

Alexeyev OA, Elgh F, Zhestkov AV, Wadell G, Juto P (1996) Hantaan and Puumala virus antibodies in blood donors in Samara, an HFRS-endemic region in European Russia. Lancet 347:9013

Alexeyev OA, Morozov VG (1995) Neurological manifestations of hemorrhagic fever with renal syndrome caused by Puumala virus: review of 811 cases. Clin Infect Dis 20:255–258

Alexeyev OA, Morozov VG, Efremov AG, Settergren B (1994) A case of hemorrhagic fever with renal syndrome complicated by spleen hemorrhage. Scand J Infect Dis 26:491–492

Antoniadis A, LeDuc JW, Daniel-Alexiou S (1987) Clinical and epidemiological aspects of hemorrhagic fever with renal syndrome (HFRS) in Greece. Eur J Epidemiol 3:295–301

Antoniadis A, Stylianakis A, Papa A, Alexiou Daniel S, Lampropoulos A, Nichol ST, Peters CJ, Spiropoulou CF (1996) Direct genetic detection of Dobrava virus in Greek and Albanian patients with hemorrhagic fever with renal syndrome. J Infect Dis 174:407–410

Arthur RR, Lofts RS, Gomez J, Glass GE, Leduc JW, Childs JE (1992) Grouping of hantaviruses by small (S) genome segment polymerase chain reaction and amplification of viral RNA from wild-caught rats. Am J Trop Med Hyg 47:210–224

Avšic-Zupanc T, Petrovec M, Furlan P, Kapš R, Elgh F, Lundkvist Å (1999) Hemorrhagic fever with renal syndrome in the Dolenjska region of Slovenia – a 10-year survey. Clin Infect Dis 28:860–865

Avšic-Zupanc T, Poljak M, Furlan P, Kapš R, Xiao SY, LeDuc JW (1994) Isolation of a strain of Hantaan virus from a fatal case of hemorrhagic fever with renal syndrome in Slovenia. Am J Trop Med Hyg 51:393–400

Avšic-Zupanc T, Xiao SY, Stojanovic R, Gligic A, van der Groen G, LeDuc JW (1992) Characterization of Dobrava virus: a hantavirus from Slovenia, Yugoslavia. J Med Virol 38:132–137

Brummer-Korvenkontio M, Henttonen H, Vaheri A (1982) Hemorrhagic fever with renal syndrome in Finland: ecology and virology of nephropathia epidemica. Scand J Infect Dis Suppl 36:88–91

Brummer-Korvenkontio M, Vaheri A, Hovi T, von Bonsdorff CH, Vuorimies J, Manni T, Penttinen K, Oker-Blom N, Lähdevirta J (1980) Nephropathia epidemica: detection of antigen in bank voles and serologic diagnosis of human infection. J Infect Dis 141:131–134

Casals J, Henderson BE, Hoogstral H, Johnson KM, Shelokov A (1970) A review of Soviet viral hemorrhagic fevers, 1969. J Infect Dis 122:437–453

Chen HX, Qui FX, Dong BJ, Ji SZ, Li YT, Wang Y, Wang HM, Zuo GF, Tao XX, Gao SY (1986) Epidemiological studies on hemorrhagic fever with renal syndrome in China. J Infect Dis 154:394–398

Childs JE, Glass GE, Korch GW, Arthur RR, Shah KV, Glasser D, Rossi C, LeDuc JW (1988) Evidence of human infection with a rat-associated hantavirus in Baltimore, Maryland. Am J Epidemiol 127:875–878

Chu YK, Jennings G, Schmaljohn A, Elgh F, Hjelle B, Lee HW, Jenison S, Ksiazek T, Peters CJ, Rollin P (1995) Cross-neutralization of hantaviruses with immune sera from experimentally infected animals and from hemorrhagic fever with renal syndrome and hantavirus pulmonary syndrome patients. J Infect Dis 172:1581–1584

Chu YK, Rossi C, LeDuc JW, Lee HW, Schmaljohn CS, Dalrymple JM (1994) Serological relationships among viruses in the *Hantavirus* genus, family *Bunyaviridae*. Virology 198:196–204

Clement J, Colson P, McKenna P (1994a) Hantavirus pulmonary syndrome in New England and Europe. N Engl J Med 331:545–546

Clement J, McKenna P, Avšic-Zupanc T, Skinner CR (1994b) Rat-transmitted hantavirus disease in Sarajevo. Lancet 344:8915

Dai ZY, Xu ZY, Wang JR, Yang JR, Yang PZ, Xu ZY (1981) Epidemiologic and clinical studies of epidemic hemorrhagic fever. Chin Med J 94:143–148

Elgh F, Linderholm M, Wadell G, Juto P (1996) The clinical usefulness of a Puumalavirus recombinant nucleocapsid protein based enzyme-linked immunosorbent assay in the diagnosis of nephropathia epidemica as compared with an immunofluorescence assay. Clinical and Diagnostic Virology 6:17–26

Elgh F, Linderholm M, Wadell G, Tärnvik A, Juto P (1998) Development of humoral cross-reactivity to the nucleocapsid protein of heterologous hantaviruses in nephropathia epidemica. FEMS Immunol Med Microbiol 22:309–315

Elgh F, Lundkvist Å, Alexeyev OA, Stenlund H, Avšic-Zupanc T, Hjelle B, Lee HW, Smith KJ, Vainionpää R, Wiger D, Wadell G, Juto P (1997) Serological diagnosis of hantavirus infections

by an enzyme-linked immunosorbent assay based on detection of immunoglobulin G and M responses to recombinant nucleocapsid proteins of five viral serotypes. J Clin Microbiol 35:1122–1130
Elgh F, Wadell G, Juto P (1995) Comparison of the kinetics of Puumala virus specific IgM and IgG antibody responses in nephropathia epidemica as measured by a recombinant antigen-based enzyme-linked immunosorbent assay and an immunofluorescence test. J Med Virol 45:146–150
Esselink RA, Gerding MN, Brouwers PJ, Solleveld H, Jordans JG, Groen J, Osterhaus ADME (1994) Guillain-Barré syndrome associated with hantavirus infection. Lancet 343:180–181
Evans TJ, Moyes D, Carpenter A, Martin R, Loetscher H, Lesslauer W, Cohen J (1994) Protective effect of 55- but not 75-kF soluble tumor necrosis factor receptor-immunoglobulin G fusion proteins in an animal model of gram-negative sepsis. J Exp Med 180:2173–2179
Forslund T, Saltevo J, Anttinen J, Auvinen S, Brummer-Korvenkontio M, Korhonen A, Poutiainen M (1992) Complications of nephropathia epidemica: three cases. J Intern Med 232:87–90
French GR, Foulke RS, Brand OA, Eddy GA, Lee HW, Lee PW (1981) Korean hemorrhagic fever: propagation of the etiologic agent in a cell line of human origin. Science 211:1046–1048
Froeb HF, McDowell ME (1954) Renal function in epidemic hemorrhagic fever. Am J Med 16:671–676
Gajdusek DC (1953) Acute infectious hemorrhagic fevers and mycotoxicoses in the Union of Soviet Socialist Republics. Medical Science Publication 2. Washington DC, Army Medical Service Graduate School, Walter Reed Army Medical Center
Gajdusek DC (1962) Virus hemorrhagic fevers – special reference to hemorrhagic fever with renal syndrome (epidemic hemorrhagic fever). J Pediatr 60:841–857
Gajdusek DC (1982) Muroid virus nephropathies and muroid viruses of the Hantaan virus group. Scand J Infect Dis 36:96–108
Gavrilovskaya IN, Apekina NS, Gorbachkova EA, Myasnikov YA, Rylceva EV, Chumakov MP (1981) Detection by enzyme-linked immunosorbent assay of haemorrhagic fever with renal syndrome virus in lung tissue of rodents from European USSR. Lancet 1:1050
Giles RB, Sheedy JA, Ekman CN, Froeb HF, Conley CC, Stockard JL, Cugell DW, Vester JW, Kiayasu RK, Entwisle G, Yoe RH (1954) The sequelae of epidemic hemorrhagic fever with a note on causes of death. Am J Med 16:629–638
Glass GE, Watson AJ, LeDuc JW, Childs JE (1994) Domestic cases of hemorrhagic fever with renal syndrome in the United States. Nephron 68:48–51
Glass GE, Watson AJ, LeDuc JW, Kelen GD, Quinn TC, Childs JE (1993) Infection with a rat-borne hantavirus in US residents is consistently associated with hypertensive renal disease. J Infect Dis 167:614–620
Gui XE, Ho M, Cohen MS, Wang QL, Huang HP, Xie QX (1987) Hemorrhagic fever with renal syndrome: treatment with recombinant alpha interferon. J Infect Dis 155:1047–1051
Hjelle B, Goade D, Torrez-Martinez N, Lang-Williams M, Kim J, Harris RL, Rawlings JA (1996) Hantavirus pulmonary syndrome, renal insufficiency, and myositis associated with infection by Bayou hantavirus. Clin Infect Dis 23:495–500
Hörling J, Lundkvist Å, Huggins JW, Niklasson B (1992) Antibodies to Puumala virus in humans determined by neutralization test. J Virol Methods 39:139–147
Hörling J, Lundkvist Å, Persson K, Mullaart M, Dzagurova T, Dekonenko A, Tkachenko E, Niklasson B (1995) Detection and subsequent sequencing of Puumala virus from human specimens by PCR. J Clin Microbiol 33:277–282
Hortling H (1946) En epidemi av fältfeber (?) i finska Lappland. Nord Med 30:1001–1004
Huggins JW, Hsiang CM, Cosgriff TM, Guang MY, Smith JI, Wu ZO, LeDuc JW, Zheng ZM, Meegan JM, Wang QN, Oland DD, Gui XE, Gibbs PH, Yuan GH, Zhang TM (1991) Prospective, double-blind, concurrent, placebo-controlled clinical trial of intravenous ribavirin therapy of hemorrhagic fever with renal syndrome. J Infect Dis 164:1119–1127
Juto P, Elgh F, Ahlm C, Alexeyev OA, Edlund K, Lundkvist Å, Wadell G (1997) The first human isolate of Puumala virus in Scandinavia as cultured from phytohemagglutinin stimulated leucocytes. J Med Virol 53:150–156
Kallio-Kokko H, Vapalahti O, Hedman K, Brummer-Korvenkontio M, Vaheri A (1993) Puumala virus antibody and immunoglobulin G avidity assays based on a recombinant nucleocapsid antigen. J Clin Microbiol 31:677–680
Kanerva M, Paakkala A, Mustonen J, Paakkala T, Lahtela J, Pasternack A (1996) Pulmonary involvement in nephropathia epidemica: radiological findings and their clinical correlations. Clin Nephrol 46:369–378
Khan AS, Spiropoulou CF, Morzunov S, Zaki SR, Kohn MA, Nawas SR, McFarland L, Nichol ST (1995) Fatal illness associated with a new hantavirus in Louisiana. J Med Virol 46:281–286

Kim YS, Ahn C, Han JS, Kim S, Lee JS, Lee PW (1995) Hemorrhagic fever with renal syndrome caused by the Seoul virus. Nephron 71:419–427

Kirsi JJ, North JA, McKernan PA, Murray BK, Canonico PG, Huggins JW, Srivastava PC, Robins RK (1983) Broadspectrum antiviral activity of 2-beta-Dribofuranosylenazole-4-carboxamide, a new antiviral agent. Antimicrob Agents Chemother 24:353–361

Kitsutani PT, Denton RW, Fritz CL, Murray RA, Todd RL, Pape WJ, Wyatt Frampton J, Young JC, Khan AS, Peters CJ, Ksiazek TG (1999) Acute Sin Nombre hantavirus infection without pulmonary syndrome, United States. Emerging Infectious Diseases 5:701–705

Kleinknecht D, Rollin PE (1992) Hypertension after hemorrhagic fever with renal syndrome. Nephron 61:121

Knight DM, Trinh H, Le J (1993) Construction and initial characterisation of a mouse-human chimaeric anti-TNF antibody. Mol Immunol 30:1443

Lähdevirta J (1971) Nephropathia epidemica in Finland. A clinical histological and epidemiological study. Ann Clin Res 3:1–54

LeDuc JW, Smith GA, Childs JE (1986) Global survey of antibody to Hantaan-related viruses among peridomestic rodents. Bull World Health Organ 64:139–144

Lee HW (1982) Korean hemorrhagic fever. Prog Med Virol 28:96–113

Lee HW (1989) Hemorrhagic fever with renal syndrome in Korea. Rev Infect Dis 11:S864–S876

Lee HW, Baek LJ, Johnson KM (1982a) Isolation of Hantaan virus, the etiologic agent of Korean hemorrhagic fever, from wild urban rats. J Infect Dis 146:638–644

Lee HW, Lee PW (1976) Korean hemorrhagic fever. I. Demonstration of causative antigen and antibodies. Korean J Intern Med 19:371–383

Lee HW, Lee PW (1977) Korean hemorrhagic fever. II. Isolation of the etiologic agent. Korean J Virol 7:19–29

Lee HW, Lee PW, Baek LJ, Chu YK (1990) Geographical distribution of hemorrhagic fever with renal syndrome and hantaviruses. Arch Virol Suppl 1:5–18

Lee HW, Lee PW, Johnson KM (1978) Isolation of the etiologic agent of Korean hemorrhagic fever. J Infect Dis 137:298–308

Lee HW, Lee PW, Lähdevirta J, Brummer-Korvenkontio M (1979) Aetiological relation between Korean haemorrhagic fever and nephropathia epidemica. Lancet 1:186–187

Lee JS (1991) Clinical features of hemorrhagic fever with renal syndrome in Korea. Kidney Int Suppl 35:S88–S93

Lee PW, Gibbs Jr CJ, Gajdusek DC, Yanagihara R (1985) Serotypic classification of hantaviruses by indirect immunofluorescent antibody and plaque reduction neutralization tests. J Clin Microbiol 22:940–944

Lee PW, Goldgaber D, Gibbs Jr CJ, Gajdusek DC, Yanagihara RT, Svedmyr A, Hlaca D, Vesenjak Hirjan J, Gligic A (1982b) Other serotypes of hemorrhagic fever with renal syndrome viruses in Europe. Lancet 2:1405–1406

Li XZ, Jim XM, Li H (1984) Semi-thin sections and electron microscopy in epidemic hemorrhagic fever. Transactions of Harbin Medical University 3:43–46

Li YL, Ruo SL, Tong Z, Ma QR, Liu ZL, Ye KL, Zhu ZY, McCormick JB, Fisher-Hoch SP, Xu ZY (1995) A serotypic study of hemorrhagic fever with renal syndrome in rural China. Am J Trop Med Hyg 52:247–251

Linderholm M, Ahlm C, Settergren B, Waage A, Tärnvik A (1996) Elevated plasma levels of tumor necrosis factor (TNF)-alpha, soluble TNF receptors, interleukin (IL)-6, and IL-10 in patients with hemorrhagic fever with renal syndrome. J Infect Dis 173:38–43

Linderholm M, Billstrom Å, Settergren B, Tärnvik A (1992) Pulmonary involvement in nephropathia epidemica as demonstrated by computed tomography. Infection 20:263–266

Linderholm M, Bjermer L, Juto P, Roos G, Sandström T, Settergren B, Tärnvik A (1993) Local host response in the lower respiratory tract in nephropathia epidemica. Scand J Infect Dis 25:639–646

Linderholm M, Sandström T, Rinnström O, Groth S, Blomberg A, Tärnvik A (1997) Impaired pulmonary function in patients with hemorrhagic fever with renal syndrome. Clin Infect Dis 25: 1084–1089

Linderholm M, Settergren B, Ahlm C, Burman L-Å, Träff S, Bäcklund U, Juto P (1991) A Swedish fatal case of nephropathia epidemica. Scand J Infect Dis 23:501–502

Lukes RJ (1954) The pathology of thirty-nine fatal cases of epidemic hemorrhagic fever. Am J Med 16:639–650

Lundkvist Å, Apekina N, Myasnikov Y, Vapalahti O, Vaheri A, Plyusnin A (1997a) Dobrava hantavirus outbreak in Russia. Lancet 350:781–782

Lundkvist Å, Hukic M, Hörling J, Gilljam M, Nichol S, Niklasson B (1997b) Puumala and Dobrava viruses cause hemorrhagic fever with renal syndrome in Bosnia-Herzegovina: evidence of highly cross-neutralizing antibody responses in early patient sera. J Med Virol 53:51–59

Lundkvist Å, Vasilenko V, Golovljova I, Plyusnin A, Vaheri A (1998) Human Dobrava hantavirus infections in Estonia. Lancet 352:9125

Meisel H, Lundkvist Å, Gantzer K, Bar W, Sibold C, Krüger DH (1998) First case of infection with hantavirus Dobrava in Germany. Eur J Clin Microbiol Infect Dis 17:884–885

Moreira AL, Sampaio EP, Zmudzinas A, Frindt P, Smith KA, Kaplan G (1993) Thalidomide exerts its inhibitory action on tumour necrosis factor alpha by enhancing mRNA degradation. J Exp Med 177:1675–1680

Mori M, Rothman AL, Kurane I, Montoya JM, Nolte KB, Norman JE, Waite DC, Koster FT, Ennis FA (1999) High levels of cytokine-producing cells in the lung tissues of patients with fatal hantavirus pulmonary syndrome. J Infect Dis 179:295–302

Morii M, Yoshimatsu K, Arikawa J, Zhou G, Kariwa H, Takashima I (1998) Antigenic characterization of Hantaan and Seoul virus nucleocapsid proteins expressed by recombinant baculovirus: application of a truncated protein, lacking an antigenic region common to the two viruses, as a serotyping antigen. J Clin Microbiol 36:2514–2521

Mustonen J, Brummer-Korvenkontio M, Hedman K, Pasternack A, Pietilä K, Vaheri A (1994a) Nephropathia epidemica in Finland: a retrospective study of 126 cases. Scand J Infect Dis 26:7–13

Mustonen J, Helin H, Pietilä K, Brummer-Korvenkontio M, Hedman K, Vaheri A, Pasternack A (1994b) Renal biopsy findings and clinicopathologic correlations in nephropathia epidemica. Clin Nephrol 41:121–126

Myhrman G (1934) En njursjukdom med egenartad symptombild. Nord Med 7:793–794

Myhrman G (1945) En ny infektionssjukdom i Nordsverige och Nordfinland. Nord Med 28:2571–2572

Myhrman G (1951) Nephropathia epidemica, a new infectious disease in Northern Scandinavia. Acta Med Scand 140:52–56

Nichol ST, Spiropoulou CF, Morzunov S, Rollin PE, Ksiazek TG, Feldmann H, Sanchez A, Childs J, Zaki S, Peters CJ (1993) Genetic identification of a hantavirus associated with an outbreak of acute respiratory illness. Science 262:914–917

Niklasson B, Jonsson M, Lundkvist Å, Hörling J, Tkachenko E (1991) Comparison of European isolates of viruses causing hemorrhagic fever with renal syndrome by a neutralization test. Am J Trop Med Hyg 45:660–665

Niklasson B, LeDuc J (1984) Isolation of the nephropathia epidemica agent in Sweden. Lancet 1: 1012–1013

Nyström K (1977) Incidence and prevalence of endemic benign (epidemic) nephropathy in AC county, Sweden, in relation to population density and prevalence of small rodents. Acta Med Scand 609:1–92

Papa A, Johnson AM, Stockton PC, Bowen MD, Spiropoulou CF, Alexiou-Daniel S, Ksiazek TG, Nichol ST, Antoniadis A (1998) Retrospective serological and genetic study of the distribution of hantaviruses in Greece. J Med Virol 55:321–327

Plyusnin A, Hörling J, Kanerva M, Mustonen J, Cheng Y, Partanen J, Vapalahti O, Kukkonen SK, Niemimaa J, Henttonen H, Niklasson B, Lundkvist Å, Vaheri A (1997) Puumala hantavirus genome in patients with nephropathia epidemica: correlation of PCR positivity with HLA haplotype and link to viral sequences in local rodents. J Clin Microbiol 35:1090–1096

Powell GM (1954) Hemorrhagic fever: a study of 300 cases. Medicine 33:99–153

Qian DY, Ding YS, Chen GF, Ding JJ, Chen YX, Lu TF, Wang ZX, Smego Jr RA (1990) A placebo-controlled clinical trial of prednisone in the treatment of early hemorrhagic fever. J Infect Dis 162:1213–1214

Rossi CA, Schmaljohn CS, Meegan JM, LeDuc JW (1990) Diagnostic potential of a baculovirus-expressed nucleocapsid protein for hantaviruses. Arch Virol Suppl 1:19–28

Schmaljohn CS, Hasty SE, Dalrymple JM, LeDuc JW, Lee HW, von Bonsdorff CH, Brummer-Korvenkontio M, Vaheri A, Tsai TF, Regnery HL, Goldgaber D, Lee PW (1985) Antigenic and genetic properties of viruses linked to hemorrhagic fever with renal syndrome. Science 227:1041–1044

Settergren B, Boman J, Linderholm M, Wiström J, Hägg E, Arvidsson PA (1992) A case of nephropathia epidemica associated with panhypopituitarism and nephrotic syndrome. Nephron 61:234–235

Settergren B, Juto P, Trollfors B, Wadell G, Norrby SR (1989) Clinical characteristics of nephropathia epidemica in Sweden: prospective study of 74 cases. Rev Infect Dis 11:921–927

Settergren B, Norrby R, Näslund U, Wadell G (1983) Case of epidemic nephropathy associated with disseminated intravascular coagulation. Lancet 2:1419

Settergren B, Stegmayr BG, Elgh F (1998) Lack of serologic evidence for an association between Hantavirus infection and end-stage renal disease. Nephron 80:1

Sheedy JA, Froeb HF, Batson HA, Conley CC, Murphy JP, Hunter RB, Cugell DW, Giles RB, Bershadsky SC, Vester JW, Yoe RH (1954) The clinical course of epidemic hemorrhagic fever. Am J Med 16:619–628

Smadel JE (1953) Epidemic hemorrhagic fever. Am J Public Health 43:1327–1330

Steer A (1955) Pathology of hemorrhagic fever. Am J Pathol 31:201–221

Svedmyr A, Lee HW, Berglund A, Hoorn B, Nyström K, Gajdusek DC (1979) Epidemic nephropathy in Scandinavia is related to Korean haemorrhagic fever. Lancet 1:100

Svedmyr A, Lee PW, Gajdusek DC, Gibbs Jr CJ, Nyström K (1980) Antigenic differentiation of the viruses causing Korean haemorrhagic fever and epidemic (endemic) nephropathy of Scandinavia. Lancet 2:315–316

Temonen M, Mustonen J, Helin H, Pasternack A, Vaheri A, Holthofer H (1996) Cytokines, adhesion molecules, and cellular infiltration in nephropathia epidemica kidneys: an immunohistochemical study. Clin Immunol Immunopathol 78:47–55

Tkachenko EA, Donets MA, Rezapkin GV, Dzagurova TK, Ivanov AP, Leshchinskaya EV, Reshetnikov I, Drozdov SG, Slonova RA, Somov GP (1982) Serotypes of HFRS (haemorrhagic fever with renal syndrome) virus in East European and Far Eastern USSR. Lancet 1:863

Tkachenko EA, Ivanov AP, Dzagurova TK, Donets MA, Rezapkin GV, Leshchinskaya EV, Zagidulin UM, Ivanova AA, Stradukhin OV, Mustafin NM, Savinova TI (1981) Immunosorbent assays for diagnosis of haemorrhagic fever with renal syndrome. Lancet 2:257–258

Valtonen M, Kauppila M, Kotilainen P, Lähdevirta J, Svartbäck CM, Kosunen O, Nurminen J, Sarkkinen H, Brummer-Korvenkontio M (1995) Four fatal cases of nephropathia epidemica. Scand J Infect Dis 27:515–517

Vapalahti O, Lundkvist Å, Kukkonen SK, Cheng Y, Gilljam M, Kanerva M, Manni T, Pejcoch M, Niemimaa J, Kaikusalo A, Henttonen H, Vaheri A, Plyusnin A (1996) Isolation and characterization of Tula virus, a distinct serotype in the genus *Hantavirus*, family *Bunyaviridae*. J Gen Virol 77: 3063–3067

WHO (1983) Haemorrhagic fever with renal syndrome: memorandum from a WHO meeting. Bull World Health Organ 61:269–275

Yanagihara R, Goldgaber D, Lee PW, Amyx HL, Gajdusek DC, Gibbs Jr CJ, Svedmyr A (1984) Propagation of nephropathia epidemica virus in cell culture. Lancet 1:1013

Zaki SR, Greer PW, Coffield LM, Goldsmith CS, Nolte KB, Foucar K, Feddersen RM, Zumwalt RE, Miller GL, Khan AS, Rollin PE, Ksiazek TG, Nichol ST, Mahy BWJ, Peters CJ (1995) Hantavirus pulmonary syndrome. Pathogenesis of an emerging infectious disease. Am J Pathol 146:552–579

Zavasky DM, Hjelle B, Peterson MC, Denton RW, Reimer L (1999) Acute infection with Sin Nombre hantavirus without pulmonary edema. Clin Infect Dis 29:664–666

Zeier M, Zöller L, Weinreich T, Padberg-Wolf E, Andrassy K, Ritz E (1992) Severe hemorrhagic complications from infection with nephropathia epidemica strain of hantavirus. Clin Nephrol 38:190–192

Zetterholm SG (1934) Akuta nefriter simulerande akuta bukfall. Läkartidningen. 31:425–429

Zöller L, Yang S, Gött P, Bautz EK, Darai G (1993) Use of recombinant nucleocapsid proteins of the Hantaan and nephropathia epidemica serotypes of hantaviruses as immunodiagnostic antigens. J Med Virol 39:200–207

Human Immune Response, Host Genetics, and Severity of Disease

O. VAPALAHTI[1], Å. LUNDKVIST[2], and A. VAHERI[1]

1 Introduction

Hantavirus infections have an incubation period of approximately 2–4 weeks after transmission and an effective immune response has usually already developed at the onset of symptoms, including high levels of specific antibodies. This makes the diagnostics of hantavirus infections rather straightforward; usually the diagnosis or exclusion of hantavirus infection can be achieved by a single measurement of hantavirus-specific IgM antibodies. However, cross-reactivities (or the lack of them) between hantaviruses should be taken into consideration when choosing the right antigen in a given geographical area, or when interpreting the exact infecting agent. In addition to being helpful for diagnostics, the coinciding onset of symptoms with the immune response together with early presence of various cytokines,

[1] Haartman Institute, University of Helsinki, Finland
[2] Swedish Institute for Infectious Disease Control and Karolinska Intitutet, Stockholm, Sweden

inflammation markers, and activated T-cells in peripheral blood and in, e.g., infected lung endothelia, has prompted researchers to seek immunopathogenetic mechanisms for the disease. Such mechanisms are further supported by the association between certain human lymphocyte antigen (HLA) haplotypes and severity of the hantavirus disease.

2 B-Cell Responses

2.1 Kinetics of the Antibody Response, Ig Classes, and Protein Specificity

Human hantavirus infections initially induce high levels of virus-specific IgM, directed to the viral nucleocapsid protein (N) and to the two viral envelope glycoproteins, G1 and G2 (Fig. 1a; Lundkvist et al. 1993a, 1997a; Zöller et al. 1993). In most Puumala virus (PUUV)-infected, i.e., nephropathia epidemica, patients, high levels of N-specific IgA and IgG (mainly of the IgA1, IgG1, and IgG3 subclasses) can also be detected already at the onset of disease (Lundkvist et al. 1993b, 1995; Groen et al. 1994; Vapalahti et al. 1995; de Carvalho Nicacio et al. 2000). Early detection of IgM, IgG, and IgA in serum (and IgA in saliva) has been demonstrated also from Andes virus (ANDV)-infected patients (Padula et al. 2000). Although early IgG responses towards N-terminal linear epitopes of G1 are found in Sin Nombre virus (SNV) infections (Jenison et al. 1994), IgG responses to native PUUV glycoproteins are detected later than those towards N (Fig. 1b; Lundkvist et al. 1993a), a phenomenon that has also been used for diagnosing acute PUUV infections using an IgG-immunofluorescence assay (Vapalahti et al. 1995). Hantavirus infections are thought to raise life-long immunity, and PUUV-specific anti-N protein IgG, in addition to neutralizing antibodies directed against G1 and G2, have been detected in convalescents several decades after the infection (Settergren et al. 1991; Lundkvist et al. 1993a).

Several methods have been applied for hantavirus diagnostics, including immunofluorescence assays (IFA) using infected cells as antigens, enzyme immunoassays (EIA) based on either native viral proteins or recombinant N protein, agglutination tests, rapid immunochromatographic tests, immunoblotting tests, hemagglutination inhibition tests, and neutralization tests.

Although a vast majority of patients have measurable IgM levels when the first serum sample is taken, it should be noted that late IgM seroresponse occasionally occurs, at least in PUUV infections. Therefore, negative IgM results exclude hantavirus infection only at (or after) the 6th day of illness (Kallio-Kokko et al. 1998). In addition, a virus-specific IgE response has been reported for PUUV infection (Alexeyev et al. 1994), but the levels were relatively low and did not correlate to any clinical parameters.

Elevated levels of PUUV-specific IgG4, especially directed to G2, in 2-year and 10- to 20-year convalescent sera have been demonstrated and a recent finding shows

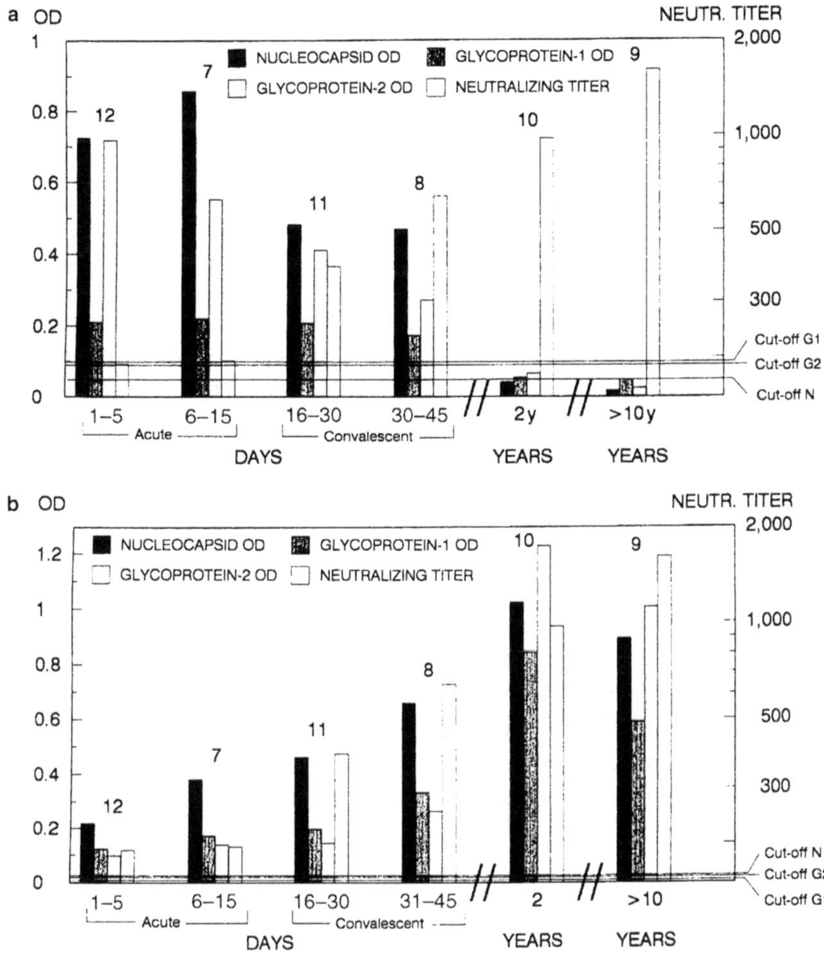

Fig. 1a,b. Acute, convalescent, and 2-year sera from 19 NE patients together with sera from 9 patients drawn 10–20 years post-NE infection were examined for **a**, IgM and **b**, IgG antibodies directed to N, G1, and G2. The bars express the mean OD at different intervals post-onset of disease. The figure above the bar shows the number of sera tested in each interval. Calculated cut-off points **a**, 0.054, 0.098, and 0.093 **b**, 0.023, 0.005, and 0.019 for the N-, G1-, and G2-enzyme-linked immunosorbent assays (ELISAs), respectively, are indicated. Taken from LUNDKVIST et al. (Arch Virol 130:121–130, 1993, pp 125)

virus-specific IgA1 in several 2-year and 10-year PUUV convalescent samples (LUNOKVIST et al. 1993b; DE CARVALHO NICACIO et al. 2000). Although virus-specific IgG4 has been demonstrated in virus infections that tend to persist in the host; e.g., hepatitis B, herpes simplex and varicella-zoster viruses (ASANO et al. 1987, 1988; MCBRIDE and WARD 1987; SÄLLBERG et al. 1990), and the role of persistent IgA as a marker for a prolonged antigenic stimulation has been suggested (FRIEDMAN and EICHLER 1991; NILSEN et al. 1991), no evidence from human patients, by reverse transcription-polymerase chain reaction (RT-PCR) or antigen detection studies, support viral persistence. For instance, PUUV RNA has been detectable only within the first days after onset of disease and not even in all patients (Hörling et al. 1995; Plyusnin et al. 1997, 1999) and PUUV antigen is generally not demonstrable in kidney biopsies in nephropathia epidemica patients (Temonen et al. 1996).

2.2 B-Cell Epitopes

2.2.1 N-protein-Specific Response

Epitope mapping has revealed the presence of several antigenic regions through the whole PUUV N protein recognized by the human antibody response (Fig. 2) (Lundkvist et al. 1995; Vapalahti et al. 1995). The recognition pattern at the epitope level varies considerably between individual patients (Vapalahti et al. 1995), which probably explains why serological assays based on short peptides or short truncated proteins have not reached optimal sensitivity (Brus Sjölander et al. 1997). In general, however, the amino-terminal 120 amino acids (aa) constitute a region reactive with most human and rodent sera (Vapalahti et al. 1995; Lundkvist et al. 1996a; Elgh et al. 1996b). Epitope mapping using monoclonal antibodies (MAbs) and truncated recombinant proteins revealed that the majority of the defined epitopes were located within the amino-terminal part of the protein: 6/8 PUUV N-epitopes were identified between aa 1–79, and 4/5 Tula virus (TULV) epitopes were identified between aa 1–61 (Fig. 2) (Lundkvist et al. 1996a,b). Similarly, the amino terminus (aa 1–103) of Hantaan virus (HTNV) N was reactive with experimentally infected mouse sera by immunoblotting and was recognized by 4/7 cross-reactive MAbs (Yoshimatsu et al. 1996). One cross-reactive epitope was

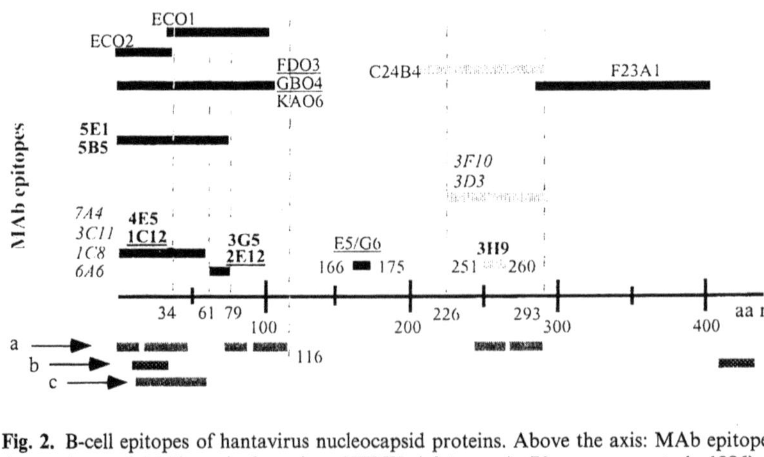

Fig. 2. B-cell epitopes of hantavirus nucleocapsid proteins. Above the axis: MAb epitopes and recognition-regions of MAbs raised against HTNV (*plain text*) (Yoshimatsu et al. 1996), PUUV (*bold*) (Lundkvist et al. 1995, 1996a), and TULV (*italics*) (Lundkvist et al. 1996b). MAbs reactive with all or most hantaviruses (*underlined*). Serotype-specific epitopes are indicated (*gray bars*). Below the axis: **a** major human B-cell epitopes to PUU (Vapalahti et al. 1995), **b** acute-phase epitopes (Lundkvist et al. 1995), and **c** major antigenic region reactive in immunoblot with sera of SNV patients (Jenison et al. 1994)

pinpointed by epitope scanning to aa 166–175 and virus-specific MAb epitopes have been defined to the most variable region (for PUUV, aa 251–260; LUNDKVIST et al. 1995) (Fig. 2). The more serotype-specific epitopes in this region have also been utilized in producing antigen to differentiate between seroresponses to the relatively closely related Seoul virus (SEOV) and HTNV viruses (MORII et al. 1998).

2.2.2 G1- and G2-Specific Response

The three-dimensional structures of the glycoproteins are complex and most MAb epitopes seem to be conformational. Neutralization and hemagglutination antibody-inducing epitopes have been found in both G1 and G2; most of the epitopes involved in hemagglutination were found in G2 (DANTAS et al. 1986; ARIKAWA et al. 1989; LUNDKVIST et al. 1992, 1993c). There were, however, many epitopes identified that induced HA but not NT activity (DANTAS et al. 1986; ARIKAWA et al. 1989). There is very limited knowledge available about the localization of virus-neutralizing epitopes, probably because of their conformational nature or dependence on glycosylation. Competition experiments with virus-neutralizing MAbs suggested the presence of different neutralizing epitopes in both G1 and G2 (ARIKAWA et al. 1989; LUNDKVIST et al. 1992, 1993c), which sometimes differ in their in vitro and in vivo neutralizing activity (ARIKAWA et al. 1992). Recently, induction of virus-neutralizing antibodies in Balb/c mice was achieved by DNA vaccination with different partially overlapping segments of SNV-G1 (aa 1–167, 265–432, 399–565, 486–652) and SNV-G2 (1–167, 134–300) (BHARADWAJ et al. 1999). Other epitope mapping approaches have been based on the analysis of virus immune escape mutants and truncated glycoprotein derivatives. For HTNV glycoproteins, epitopes of non-neutralizing MAbs have been mapped to the N-terminal third of HTNV G1, by using truncated recombinant antigens (WANG et al. 1993). For SNV, the G1 N-terminus (aa 59–89) has also been shown to contain a linear, serotype-specific human B-cell epitope (JENISON et al. 1994).

Analysis of HTNV neutralizing MAb escape mutants (for three G1 and two G2-specific neutralizing MAbs) demonstrated in some cases single nucleotide exchanges in one codon, while some mutants contained substitutions at distantly located positions. The analysis of the reactivity of C-terminally truncated G2 derivatives revealed that certain virus-neutralizing G2-specific MAbs probably require two distantly located regions for antigen binding. However, deleting more than 20% of the total G2 protein resulted in loss of reactivity with MAbs. Epitopes of non-neutralizing G1 MAbs were mapped to the amino-terminal third of HTNV G1 (WANG et al. 1993). The analysis of two MAb escape mutants, rescued from virus-neutralizing MAbs against PUUV G1 and G2, revealed only single amino acid exchanges. However, peptides covering the corresponding wild-type sequences were not recognized by the MAbs and did not inhibit the antigen binding of the MAbs; anti-peptide sera did not recognize the native viral proteins. Therefore, it was suggested that these epitopes were non-linear or glycosylated in the native protein, or alternatively, that the MAbs reacted with regions away from the mutations (HÖRLING and LUNDKVIST 1997). HEISKANEN et al. (1999), utilizing a random 18-amino acid peptide library that was displayed on filamentous phages and peptides synthesized on membranes, identified peptides that could be aligned with primary sequences of PUUV G1 and G2. Similarities of the selected phage clones to a MAb-escape mutant site (MAb 5A2, D272 → V; HÖRLING and LUNDKVIST 1997) were found.

No serological responses to the large polymerase (L) protein have been described.

2.3 Neutralizing Activity

The neutralizing antibody response develops early and is usually present already at the onset of disease (HÖRLING et al. 1992; LUNDKVIST et al. 1993a, 1997a). Recent data have revealed a high degree of hantavirus-serotype cross-reactivity of the neutralizing antibodies in HFRS acute-phase serum samples (LUNDKVIST et al. 1997a). Since only low levels of IgG are present towards the viral glycoproteins at the onset of disease (Fig. 1b; LUNDKVIST et al. 1993a), these results indicate the presence of neutralizing, broad-reactive IgM. Further, neutralizing IgA1 has been detected in some acute-phase PUUV patients (DE CARVALHO NICACIO et al. 2000), but the degree of cross-reactions to other hantaviruses is still unknown. This complex cross-reactivity of the early neutralizing antibodies, which varies remarkably among individuals, revealed that sera drawn within the first month after onset of disease should not be used for serotyping (examples are given in Table 1). In contrast, in late HFRS convalescent sera, the neutralizing antibodies are highly serotype-specific and several recent studies have revealed the usefulness

Table 1. Examples of FRNT analyses of sequentially drawn sera from six HFRS patients illustrating high levels of serotype cross-neutralization during the first month after onset of disease. (From LUNDKVIST et al. 1997a)

Patient-sample (time)[a]	Virus			
	PUU	HTN	DOB	SEO
I-1 (10d)	80[b]	160	20	20
I-2 (12m)	320	<20	20	20
2-I (7d)	80	640	40	<20
2-II (11m)	320	40	20	<20
3-I (18d)	40	160	80	80
3-II (25d)	80	160	320	20
3-III (8m)	20	80	640	80
4-I (12d)	80	80	20	<20
4-II (15m)	320	40	20	<20
5-I (11d)	160	160	20	<20
5-II (15m)	1280	40	20	<20
6-I (9d)	160	160	20	20
6-II (12m)	160	20	<20	<20

[a] Days (d) or months (m) after onset of disease for sampling;
[b] FRNT end-point titers.

of the neutralization assay in epidemiological studies (LUNDKVIST et al. 1997a,b, 1998a,b; MEISEL et al. 1998; AVSIC-ZUPANC et al. 1999; HEYMAN et al. 1999; SIBOLD et al. 1999). Extended studies are needed to better understand the kinetics of the humoral response in HFRS patients, e.g., how the different immunoglobulin classes are involved in virus neutralization and what the molecular basis is for the complex cross-reactions.

The viral envelope glycoproteins are presumed to be the major elements involved in induction of immunity to hantaviruses. This assumption is based on passive protection experiments and by the neutralizing activity detected in vitro for MAbs directed to G1 and G2, but not to N (DANTAS et al. 1986; ARIKAWA et al. 1989, 1992; ZHANG et al. 1989; SCHMALJOHN et al. 1990; LUNDKVIST and NIKLASSON 1992). The significance of the N-specific antibody response in vivo is, however, not yet completely understood. A MAb specifically for HTNV N has been shown to protect from virus infection and N-specific polyclonal sera to significantly increase the time before death in a mouse model (YOSHIMATSU et al. 1993). N-specific MAbs have been shown to partially protect bank voles from PUUV infection (Lundkvist et al., unpublished observations). Accordingly, the humoral response to N may, in addition to the glycoprotein-specific response, be of importance for the immunity, e.g., via antibody-dependent cell-mediated cytotoxicity or complement-mediated cytolysis.

Less is known about the antibody responses in the rodent hosts. It is believed that the IgG response persists lifelong. Usually, infected rodents show high titers of virus-specific IgG in sera (BERNSHTEIN et al. 1999), and significant levels of highly serotype-specific neutralizing antibodies are found in wild-trapped as well as in experimentally infected rodents, which makes the neutralization assay a valuable tool for the typing of the infecting agent (LUNDKVIST et al. 1997a, 1998b; ESCUTENAIRE et al. 2000). In bank voles, maternal PUUV-specific antibodies have been detected in the offspring, which may temporarily protect them from infection (GAVRILOVSKAYA et al. 1990; BERNSHTEIN et al. 1999; our unpublished observations). In rats, maternal SEOV-specific antibodies have been shown to be vertically transferred in utero and by milk (DOHMAE et al. 1993; DOHMAE and NISHIMUNE 1998).

Recent studies (BHARADWAJ et al. 2000) have shown that patients who were destined to survive from hantavirus cardiopulmonary syndrome have higher neutralizing antibody titers to SNV compared with those who did not.

2.4 Serological Cross-Reactivities Between Hantaviruses

The humoral cross-reactivity between the different hantavirus serotypes is high, especially within the group of viruses carried by (1) Murinae rodents (HTNV, SEOV and DOBV), (2) Arvicolinae rodents (PUUV, PHV, TULV and TOPV), and (3) Sigmodontinae rodents (SNV- and ANDV-like viruses; CHU et al. 1994). The N protein amino acid homologies between the first and the latter two groups are in the range of 58–63%, between the two latter groups 70–76% and within the groups >79%; the serological cross-reactivities mirror these differences. This has resulted in the general impression that only two antigens (HTNV and PUUV) or one antigen (SNV) should be enough for the efficient serological diagnosis of hantavirus infections in Eurasia and in the Americas, respectively.

While the sera cross-react with heterologous N antigens, especially within the three aforementioned serological groups, but also between the different groups, e.g., antibodies produced in PUUV patients in most cases react with TULV or TOPV antigens and HPS patient sera usually react with PUUV N (our unpublished observation; Jenison et al. 1994), early sera are not necessarily optimally recognized by heterologous antigens. For example, the recent awareness of DOBV as an important human pathogen in large parts of Europe (Avsic-Zupanc et al. 1995; Antoniadis et al. 1996; Lundkvist et al. 1997a; Nemirov et al. 1999) prompted the development and evaluation of DOBV-specific serological assays (Brus Sjölander and Lundkvist 1999; Kallio-Kokko et al. 2000). The results confirmed the high degree of antibody cross-reactivity between DOBV and HTNV antigen-based EIAs, but also proved the usefulness of homologous antigen for diagnosis/sero-epidemiology of DOBV infections; e.g., 7.1% of the DOBV-positive acute-phase patient samples and 12.5% of the DOBV convalescent samples were found negative when examined for specific IgG against the heterologous HTNV antigen. Similar investigations concerning SNV- and ANDV-related viruses in the Americas are in progress.

2.5 Comparison of Assays for Hantavirus Antibody Detection

Since recent studies have shown that complete N protein is a more sensitive antigen than its short amino-terminal fragments (Brus Sjölander et al. 1997; Kallio-Kokko et al. 2000), assays utilized for sero-epidemiological screening, as well as assays for patient diagnosis, should be based on the complete N protein. During recent years, a number of recombinant hantavirus antigens have been introduced. Comparable studies using MAbs revealed that all investigated *E. coli*-expressed PUUV N recombinants lacked one or two out of eight B-cell epitopes present in the native protein (Elgh et al. 1996b; Vapalahti et al. 1996; Lundkvist et al. 1996a). Also, bacterially produced recombinant SNV N lacks some N-terminal MAb epitopes (Lundkvist et al. 1996b; Vapalahti et al. 1999). In contrast, HTNV, PUUV, and DOBV N proteins expressed in insect cells have been found indistinguishable from the native proteins (Schmaljohn et al. 1990; Vapalahti et al. 1996; Kallio-Kokko et al. 2000; Brus Sjölander et al. 2000). The use of *E. coli*-expressed viral antigens may be further hampered by non-specific reactions to *E. coli*-derived contaminants; despite extensive purification of the recombinant antigen, adsorption of serum samples to *E. coli*-extracts has been required to minimize non-specific reactions (Elgh et al. 1996a; Brus Sjölander et al. 1997). In addition, EIAs based on direct coating of crude preparations of native or baculovirus-expressed viral antigens usually suffer from non-specific reactions. To avoid problems with non-specific antibody reactions, the use of a capture format, in which the antigen is captured by MAbs, seems to be the optimal strategy (K. Brus Sjölander et al. 1997, unpublished observations). However, recent rapid immunochromatographic tests with highly purified recombinant antigens have also shown promising results for "point-of-care" use (our unpublished results).

Although EIA is optimal for a highly specific serological confirmation of hantavirus infections, e.g., for patient diagnosis, the antibody responses usually cross-react strongly between, for example, PUUV, DOBV, SEOV, and HTNV antigens (ELGH et al. 1997; LUNDKVIST et al. 1997a). This revealed that EIA, or other rapid assays such as IFA or immunoblotting, which mainly detect antibodies directed against the immunodominant, cross-reactive N protein, could generally not be used for serotyping, although assays to detect only SNV antibodies and to distinguish between SEOV and HTNV infections have been reported (JENISON et al. 1994; MORII et al. 1998). The exact infecting hantavirus type can, therefore, only be determined either by the time- and labor-consuming neutralization assay comparing serum titers to all relevant hantaviruses, or by RT-PCR amplification of hantavirus RNA followed by sequencing.

3 T-Cell and Cytokine Responses

3.1 Cellular Immune Response

Little is known about the cellular immune responses to hantaviruses, but early reports have suggested that a cell-mediated response to HTNV is involved in protection (ASADA et al. 1987, 1988; YOSHIMATSU et al. 1993). An increase in the number of activated circulating $CD8^+$ cells expressing HLA DR and CD25 antigens has been described in HTNV-caused HFRS during illness, along with an increase in the number of activated T cells producing interferon (IFN)-γ (HUANG et al. 1994). A reversed $CD4^+/CD8^+$ ratio has also been documented in the early stage of HFRS (CHEN and YANG 1990). $CD8^+$ phenotype T lymphocytes together with monocyte/macrophages have been found in association with hantavirus-positive lung endothelial cells of HPS patients (ZAKI et al. 1995) and activation of $CD8^+$ cells has been demonstrated in HPS patients (NOLTE et al. 1995). Infiltrating cells near tubuli included predominantly $CD8^+$ T lymphocytes, together with plasma cells and monocyte/macrophages in PUUV patients (TEMONEN et al. 1996).

CTL epitopes were recently identified in the N and G2 proteins from acute-phase HPS patients infected with SNV (ENNIS et al. 1997), including isolation directly from blood of a $CD8^+$ clone recognizing a SNV-specific variable sequence (aa 232–242) and producing IFN-γ. Another two clones isolated were more cross-reactive (e.g., with PUUV) including a HLA B 35-restricted CD8 epitope (aa 131–139) and a CD4 epitope from the conserved C-terminal region (372–380). More recently, analysis of human memory CTL responses against laboratory acquired HTNV infection and natural PUUV infection have identified both virus-specific and cross-reactive CTL epitopes in the N protein and responses towards G1, and for natural PUUV infection, also towards G2 (VAN EPPS et al. 1999; H.L.

Van Epps et al. unpublished observations). Two HTNV CD8$^+$ cell lines recognized a HLA B51-restricted N terminal virus-specific epitope (aa 1–20) and a HLA A1-restricted epitope from the very C-terminus (aa 421–429), a sequence which is shared between almost all hantaviruses.

The N proteins of hantaviruses are highly immunogenic in both laboratory animals and humans and have also been shown to efficiently induce protective immunity in animal models (Schmaljohn et al. 1990; Lundkvist et al. 1993a, 1996a; Yoshimatsu et al. 1993; Ulrich et al. 1998). Since the humoral response to N protein is not neutralizing in vitro, the induced protection has been regarded as mainly cell-mediated.

3.2 Cytokine Responses

Increased levels of cytokines have been encountered both in vivo and in vitro. TNF-α is a cytokine the production of which could be related to symptoms observed in HFRS and HPS. Elevated levels of TNF-α, as well as soluble TNF receptors, interleukin (IL)-6 and IL-10, have been reported in HFRS patients (Linderholm et al. 1996). In some studies, elevated levels of histamine and serotonin (Sidelnikov and Sivoraksha 1990), possibly in context with increased IgE levels (Alexeyev et al. 1994) have been noted; also, elevated levels of nitric oxide metabolites have been measured (Groeneveld et al. 1995) all of which could contribute to the vascular and endothelial disturbances. Locally, in kidney biopsies of HFRS patients, increased levels of adhesion molecules, cytokines, particularly TNF-α in conjunction with infiltrating cells near the tubuli, can be found (Temonen et al. 1996). In HPS patients, high numbers of cytokine-producing cells were seen in the lung and spleen tissues, but not kidney or liver (Mori et al. 1999). A recent study illustrated a strong specific activation of monocytes in a severe HPS-like PUUV infection, not found in other PUUV infections (Takala et al. 2000). In the same study, an association of the cytokine levels (e.g., IL-6) with platelet count and blood pressure was found, but not with the degree of renal injury.

4 Association of HLA Haplotype and Cytokine Expression with the Clinical Course of Hantavirus Infections

The different hantaviruses differ drastically in their pathogenicity, ranging from apathogenic to highly virulent HPS- and HFRS-causing viruses. Moreover, the clinical course of individual hantavirus infections is extremely variable and this is especially true for HFRS-causing viruses PUUV, DOBV, HTNV, and SEOV (Kanerva et al. 1998a). In the case of DOBV hantavirus there is growing evidence that the virus of the DOBV-Slovenia genetic lineage, originally isolated from yellow-necked field mice in the Balkans, causes a more severe disease than virus of the DOBV-Saaremaa lineage, which was first isolated from a striped field mouse trapped in Estonia (NEMIROV et al. 1999; AVSIC-ZUPANC et al. 2000). In addition, there is very preliminary evidence that SNV and PUUV viruses infecting humans may at least sometimes be reassortants of two lineages of rodent-borne viruses (LI et al. 1995; PLYUSNIN et al. 1997).

The mechanisms of hantaviral pathogenesis are poorly known (KANERVA et al. 1998a). The viruses grow poorly in cell culture, cause little or no cytopathology, and no good animal models exist. The general assumption is that increased capillary permeability is a key factor, but potential mediators have not yet been defined. An important discovery was that β_3 integrins mediate the entry of pathogenic and β_1 integrins the entry of apathogenic hantaviruses to host cells (GAVRILOVSKAYA et al. 1998, 1999).

Another advance in the field was when the severity of PUUV infection was linked to certain host genes, the susceptibility MHC haplotype for severe PUUV infection. HLA alleles B8, DR3, DQ2 are strongly associated with severe outcome of PUUV infection (MUSTONEN et al. 1996); these HLA alleles B8, DR3, DQ2 have also been associated with presence of PUUV RNA in clinical samples (PLYUSNIN et al. 1997; KANERVA et al. 1998b). In contrast, the HLA B27 allele is associated with a mild course of PUUV infection (MUSTONEN et al. 1998). Whether similar genetic factors operate also in other hantavirus infections (HTNV, DOBV, SEOV, SNV, ANDV) remains to be determined. Preliminary evidence suggests the association of the allele HLA-B35 and fatal SNV HPS (MERZ et al. 1999), but the material is too small for firm conclusions.

The observation that severe PUUV infection is associated with the extended HLA haplotype containing HLA alleles B8, DR3, DQ2 suggests that inflammatory host immune response may alter the course of the clinical disease. This could be due to mechanisms by which the virus escapes the immune surveillance, or due to defective T-cell activation associated with the particular HLA haplotype. Alternatively, the HLA haplotype could be associated with an inappropriate immune response, not unlike in autoimmune diseases. Interestingly, the above risk HLA haplotype has been associated with a rapid progression to AIDS in HIV infection (STEEL et al. 1988; MALKOVSKY 1996), a poor antibody response to hepatitis B vaccine (ALPER et al. 1989; EGEA et al. 1991) and certain autoimmune diseases such as systemic lupus erythematosus, celiac disease, and Grave's disease (TIWARI and TERASAKI 1985).

It seems possible that individuals with the risk HLA alleles, intensive Th1-type cytokine responses, and a high virus load will have more vigorous virus-specific T-lymphocyte responses and to be at greater risk for severe HFRS. So far, differences have not been found in proliferative T-helper cell responses to purified PUUV, in the total anti-PUU or neutralizing antibody titers between the allele-positive and negative patients (M. Kanerva et al., unpublished observations). Study of the role of cytotoxic T-cells will be of special interest. However, the HLA alleles linked to the susceptibility may also merely act as markers of other gene(s) in linkage disequilibrium with this haplotype and indeed, certain complement genes are inherited together with this extended haplotype. Especially interesting is the C4A null allele which may lead to total or partial absence of C4A and to impaired immune complex clearance from circulation (Morgan and Walport 1991).

TNF-α is one of the most important proinflammatory cytokines. There is evidence that overproduction of TNF-α may lead to severe systemic toxicity, and it acts as a mediator of septic shock. TNF-α release from macrophages may be involved in the endothelial leakage characteristic of filovirus hemorrhagic fevers (Feldmann et al. 1996). The biallelic polymorphism in the TNF-α gene promoter region (-308) has been studied in nephropathia epidemica patients. The more rare TNF2 allele, associated with a higher TNF-α production by mononuclear cells (Wilson et al. 1997), was found to be significantly more frequent in hospitalized nephropathia epidemica patients than in healthy controls (Kanerva et al. 1998b). In support of a clinical relevance, the TNF2 allele has been associated with such infections as severe cerebral malaria, fatal meningococcal meningitis, mucocutaneous leishmaniasis and lepromatous leprosy (McGuire et al. 1994; Cabrera et al. 1995; Nadel et al. 1996; Roy et al. 1997). The genetically determined high-producer TNF-α phenotype could thus play a role in severe PUUV infection. This is an attractive hypothesis since increased serum levels (Yang and Bang 1992; Linderholm et al. 1996) and prominent expression in kidney biopsies of TNF-α (Temonen et al. 1996) have been reported in PUUV-caused HFRS. Yet, it is not known whether the TNF2 allele is an independent risk factor of severe PUUV infection or merely a passive component of the extended haplotype. According to a recent study (Mäkelä et al. 2000), non-carriers of interleukin-1 receptor antagonist (IL-1RA) allele 2 and IL-1b (-511) allele 2 may also contribute to susceptibility to NE.

5 Concluding Remarks

There is still much to learn about the immune responses to hantavirus infections. An extended knowledge on virus-host interactions will be of great importance for pathological, diagnostic, and epidemiological questions, as well as for the development of efficient therapeutic and protective reagents. The present knowledge concerning cellular immune responses is limited but growing. The recent evidence on the association of certain HLA alleles to disease severity, and the significant increases of specific cytokines as IL-6, IL-10, and TNF-α, prompt future studies on immunopathological mechanisms of hantavirus diseases.

References

Alexeyev OA, Ahlm C, Billheden J, Settergren B, Wadell G, Juto P (1994) Elevated levels of total and Puumala virus-specific immunoglobulin E in the Scandinavian type of hemorrhagic fever with renal syndrome. Clin Diagn Lab Immunol 1:269–272

Alper CA, Kruskall MS, Marcus-Bagley D, Craven DE, Katz AJ, Brink SJ, Dienstag JL, Awdeh Z, Yunis EJ (1989) Genetic prediction of nonresponse to hepatitis B vaccine. N Engl J Med 321:708–712

Antoniadis A, Stylianakis A, Papa A, Alexiou-Daniel S, Lampropoulos A, Nichol ST, Peters CJ, Spiropoulou CF (1996) Direct genetic detection of Dobrava virus in Greek and Albanian patients with hemorrhagic fever with renal syndrome. J Infect Dis 174:407–410

Arikawa J, Schmaljohn AL, Dalrymple JM, Schmaljohn CS (1989) Characterization of Hantaan virus envelope glycoprotein antigenic determinants defined by monoclonal antibodies. J Gen Virol 70: 615–624

Arikawa J, Yao J-S, Yoshimatsu K, Takashima I, Hashimoto N (1992) Protective role of antigenic sites on the envelope protein of Hantaan virus defined by monoclonal antibodies. Arch Virol 126:271–281

Asada H, Tamura M, Kondo K, Okuno Y, Takahashi Y, Dohi Y, Nagai T, Kurata T, Yamanishi K (1987) Role of T lymphocyte subsets in protection and recovery from Hantaan virus infection in mice. J Gen Virol 68:1961–1069

Asada H, Tamura M, Kondo K, Dohi Y, Yamanishi K (1988) Cell-mediated immunity to virus causing haemorrhagic fever with renal syndrome: generation of cytotoxic T lymphocytes. J Gen Virol 69:2179–2188

Asano Y, Hiroishi Y, Itakura N, Hirose S, Kajita Y, Nagai T, Yazaki T, Takahashi M (1987) Immunoglobulin subclass antibodies to varicella-zoster virus. Pediatrics 80:933–936

Asano Y, Hiroishi Y, Itakura N, Hirose S, Kajita Y, Suga S, Yazaki T (1988) Placental transfer of IgG subclass-specific antibodies to varicella-zoster virus. J Med Virol 26:1–6

Avsic-Zupanc T, Toney A, Anderson K, Chu YK, Schmaljohn C (1995) Genetic and antigenic properties of Dobrava virus: a unique member of the *Hantavirus* genus, family *Bunyaviridae*. J Gen Virol 76:2801–2808

Avsic-Zupanc T, Petrovec M, Furlan P, Kaps R, Elgh F, Lundkvist Å (1999) Hemorrhagic Fever with Renal Syndrome in the Dolenjska Region of Slovenia – A 10-Year Survey. Clin Infect Dis 28:860–865

Avsic-Zupanc T, Nemirov K, Petrovec M, Trilar T, Poljak M, Vaheri A, Plyusnin A (2000) Genetic analysis of wild-type Dobrava hantavirus in Slovenia: co-existence of two distinct genetic lineages within the same natural focus. J Gen Virol 81:1747–1755

Bernshtein AD, Apekina NS, Mikhailova TV, Myasnikov YA, Khlyap LA, Korotkov YS, Gavrilovskaya IN (1999) Dynamics of Puumala hantavirus infection in naturally infected bank voles (*Clethrinomys glareolus*). Arch Virol 144:2415–2428

Bharadwaj M, Lyons CR, Wortman IA, Hjelle B (1999) Intramuscular inoculation of Sin Nombre hantavirus cDNAs induces cellular and humoral immune response in BALBc mice. Vaccine 17: 2836–2843

Bharadwaj M, Nofchissey R, Goade D, Koster F, Hjelle B (2000) Humoral immune responses in the hantavirus cardiopulmonary syndrome. J Infect Dis 182:43–48

Brus Sjölander K, Elgh F, Kallio-Kokko H, Vapalahti O, Hägglund M, Palmcrantz V, Juto P, Vaheri A, Niklasson B, Lundkvist Å (1997) Evaluation of serological methods for diagnosis of Puumala hantavirus infection (nephropathia epidemica). J Clin Microbiol 35:3264–3268

Brus Sjölander K, Lundkvist Å (1999) Dobrava virus infection: serological diagnosis and cross-reactions to other hantaviruses. J Virol Methods 80:137–143

Brus Sjölander K, Golovljova I, Plyusnin A, Lundkvist Å (2000) Diagnostic potential of Puumala virus nucleocapsid protein expressed in *Drosophila melanogaster* cells. J Clin Microbiol 38:2324–2329

Cabrera M, Shaw MA, Sharples C, Williams H, Castes M, Convit J, Blackwell JM (1995) Polymorphism in tumor necrosis factor genes associated with mucocutaneous leishmaniasis. J Exp Med 182:1259–1264

de Carvalho Nicacio C, Björling E, Lundkvist Å (2000) Immunoglobulin A responses to Puumala hantavirus. J Gen Virol 81:1453–1461

Chen LB, Yang WS (1990) Abnormalities of T cell immunoregulation in hemorrhagic fever with renal syndrome. J Infect Dis 161:1016–1019

Chu YK, Rossi C, LeDuc JW, Lee HW, Schmaljohn CS, Dalrymple JM (1994) Serological relationships among viruses in the *Hantavirus* genus, family *Bunyaviridae*. Virology 198:196–204

Dantas JR Jr, Okuno Y, Asada H, Tamura M, Takahashi M, Tanishita O, Takahashi Y, Kurata T, Yamanishi K (1986) Characterization of glycoproteins of viruses causing hemorrhagic fever with renal syndrome (HFRS) using monoclonal antibodies. Virology 151:379–384

Dohmae K, Koshimizu U, Nishimune Y (1993) In utero and mammary transfer of hantavirus antibody from dams to infant rats. Lab Anim Sci 43:557–561

Dohmae K, Nishimune Y (1998) Maternal transfer of Hantavirus antibodies in rats. Lab Anim Sci 48:395–397

Egea E, Iglesias A, Salazar M, Morimoto C, Kruskall MS, Awdeh Z, Schlossman SF, Alper CA, Yunis EJ (1991) The cellular basis for lack of antibody response to hepatitis B vaccine in humans. J Exp Med 173:531–538

Elgh F, Linderholm M, Wadell G, Juto P (1996a) The clinical usefulness of a Puumala virus recombinant nucleocapsid protein based enzyme-linked immunosorbent assay in the diagnosis of nephropathia epidemica as compared with an immunofluorescence assay. Clin Diag Virol 6:17–26

Elgh F, Lundkvist Å, Alexeyev OA, Wadell G, Juto P (1996b) A major antigenic domain for the human humoral response to Puumala virus nucleocapsid protein is located at the amino-terminus. J Virol Methods 59:161–172

Elgh F, Lundkvist Å, Alexeyev OA, Stenlund H, Avsic-Zupanc T, Hjelle B, Lee HW, Smith KJ, Vainionpää R, Wiger D, Wadell G, Juto P (1997) Serological diagnosis of hantavirus infection by enzyme-linked immunosorbent assay based on detection of immunoglobulin G and M responses to recombinant nucleocapsid protein of five viral serotypes. J Clin Microbiol 35:1122–1130

Ennis F, Cruz J, Spiropoulou C, Waite D, Peters C, Nichol S, Kariwa H, Koster F (1997) Hantavirus pulmonary syndrome: CD8+ and CD4+ cytotoxic T lymphocytes to epitopes on Sin Nombre virus nucleocapsid protein isolated during acute illness. 238:380–390

Escutenaire S, Chalon P, Verhagen R, Heyman P, Thomas I, Karelle-Bui L, Avsic-Zupanc T, Lundkvist Å, Plyusnin A, Pastoret PP (2000) Spatial and temporal dynamics of Puumala hantavirus infection in red bank vole (*Clethrionomys glareolus*) populations in Belgium. Virus Research, in press

Feldmann H, Bugany H, Mahner F, Klenk HD, Drenckhahn D, Schnittler HJ (1996) Filovirus-induced endothelial leakage triggered by infected monocytes/macrophages. J Virol 70:2208–2214

Friedman MG, Eichler AW (1991) The use of serum IgA as a diagnostic marker in viral infections. Immunology and Infectious Diseases 1:223–234

Gavrilovskaya I, Apekina NS, Bernstein AD, Demina VD, Okulova NM, Myasnikov YA, Chumakov MP (1990) Pathogenesis of HFRS virus infection and mode of horizontal transmission of hantavirus in bank voles. Arch Virol Suppl 1:57–62

Gavrilovskaya IN, Shepley M, Shaw R, Ginsberg MH, Mackow ER (1998) β3 integrins mediate the cellular entry of hantaviruses that cause respiratory failure. Proc Natl Acad Sci USA 95:70744–7079

Gavrilovskaya IN, Brown EJ, Ginsberg MH, Mackow ER (1999) Cellular entry of hantaviruses which cause hemorrhagic fever with renal syndrome is mediated by beta3 integrins. J Virol 73:3951–3959

Groen J, Gerding M, Jordans JGM, Clement JP, Osterhaus ADME (1994) Class and subclass distribution of hantavirus-specific serum antibodies at different times after the onset of nephropathia epidemica. J Med Virol 43:39–43

Groeneveld PH, Colson P, Kwappenberg KM, Clement J (1995) Increased production of nitric oxide in patients infected with the European variant of hantavirus. Scand J Infect Dis 27:453–456

Heiskanen T, Lundkvist Å, Soliymani R, Koivunen E, Vaheri A, Lankinen H (1999) Phage-displayed peptides mimicking the discontinuous neutralization sites of puumala Hantavirus envelope glycoproteins. Virology 262:321–332

Heyman P, Vervoort T, Colson P, Chu Y K, Avsic-Zupanc T, Lundkvist Å (1999) A major outbreak of hantavirus infection in Belgium in 1995 and 1996. Epidemiol Infect 122:447–453

Hörling J, Lundkvist Å, Huggins JW, Niklasson B (1992) Antibodies to Puumala virus in humans determined by neutralization test. J Virol Methods 39:139–147

Hörling J, Lundkvist Å, Persson K, Mullaart M, Dzagurova T, Dekonenko A, Tkachenko E, Niklasson B (1995) Detection and subsequent sequencing of Puumala virus from human specimens by PCR. J Clin Microbiol 33:277–282

Hörling J, Lundkvist Å (1997) Single amino acid substitutions in Puumala virus envelope glycoproteins G1 and G2 eliminate important neutralization epitopes. Virus Research 48: 89–100

Huang C, Jin B, Wang M, Li E, Sun C (1994) Hemorrhagic fever with renal syndrome: relationship between pathogenesis and cellular immunity. J Infect Dis 169: 868–870

Jenison S, Yamada T, Morris C, Anderson B, Torrez-Martinez N, Keller N, Hjelle B (1994) Characterization of human antibody responses to four corners hantavirus infections among patients with hantavirus pulmonary syndrome. J Virol 68:3000–3006

Kallio-Kokko H, Vapalahti O, Lundkvist Å, Vaheri A (1998) Evaluation of Puumala virus IgG and IgM enzyme immunoassays based on recombinant baculovirus expressed nucleocapsid protein for early nephropathia epidemica diagnosis. Clin Diag Virol 10:83–90

Kallio-Kokko H, Lundkvist Å, Plyusnin A, Avsic-Zupanc T, Vaheri A, Vapalahti O (2000) Antigenic properties and diagnostic potential of recombinant Dobrava virus nucleocapsid protein. J Med Virol 61:266–274

Kanerva M, Mustonen J, Vaheri A (1998a) Pathogenesis of Puumala and other hantavirus infections. Rev Med Virol 8:67–86
Kanerva M, Vaheri A, Mustonen J, Partanen J (1998b) High-producer allele of tumour necrosis factor-alpha is part of the susceptibility MHC haplotype in severe Puumala virus-induced nephropathia epidemica. Scand J Infect Dis 30:532–534
Li D, Schmaljohn AL, Anderson K, Schmaljohn CS (1995) Complete nucleotide sequences of the M and S segments of two hantavirus isolates from California: evidence for reassortment in nature among viruses related to hantavirus pulmonary syndrome. Virology 206:973–983
Linderholm M, Ahlm C, Settergren B, Waage A, Tärnvik A (1996) Elevated plasma levels of tumor necrosis factor (TNF)-alpha, soluble TNF receptors, interleukin (IL)-6, and IL-10 in patients with hemorrhagic fever with renal syndrome. J Infect Dis 173:38–43
Lundkvist Å, Niklasson B (1992) Bank vole monoclonal antibodies against Puumala virus envelope glycoproteins: identification of epitopes involved in neutralization. Arch Virol 26:93–105
Lundkvist Å, Hörling J, Niklasson B (1993a) The humoral response to Puumala virus infection (nephropathia epidemica) investigated by viral protein specific immunoassays. Arch Virol 130: 121–130
Lundkvist Å, Björsten S, Niklasson B (1993b) Immunoglobulin G subclass responses against the structural components of Puumala virus. J Clin Microbiol 31:368–372
Lundkvist Å, Hörling J, Athlin L, Rosén A, Niklasson B (1993c) Neutralizing human monoclonal antibodies against Puumala virus causative agent of nephropathia epidemica: a novel method using antigen-coated magnetic beads for specific B cell isolation. J Gen Virol 74:1303–1310
Lundkvist Å, Björsten S, Niklasson B, Ahlborg N (1995) Mapping of B-cell determinants in the nucleocapsid protein of Puumala virus; definition of epitopes specific for acute IgG recognition in humans. Clin Diagn Lab Immunol 2:82–86
Lundkvist Å, Kallio-Kokko H, Brus Sjölander K, Lankinen H, Niklasson B, Vaheri A, Vapalahti O (1996a) Characterization of Puumala virus nucleocapsid protein: identification of B-cell epitopes and domains involved in protective immunity. Virology 216:397–406
Lundkvist Å, Vapalahti O, Plyusnin A, Brus Sjölander K, Niklasson B, Vaheri A (1996b) Characterization of Tula virus antigenic determinants defined by monoclonal antibodies raised against baculovirus-expressed nucleocapsid protein. Virus Research 45:29–44
Lundkvist Å, Hukic M, Hörling J, Gilljam M, Nichol S, Niklasson B (1997a) Puumala and Dobrava viruses cause hemorrhagic fever with renal syndrome in Bosnia-Herzegovina: evidence of highly cross-neutralizing antibody responses in early patient sera. J Med Virol 53:51–59
Lundkvist Å, Apekina N, Myasnikov Y, Vapalahti O, Vaheri A, Plyusnin A (1997b) Dobrava hantavirus outbreak in Russia. Lancet 350:781–782
Lundkvist Å, Vasilenko V, Golovljova I, Plyusnin A, Vaheri A (1998a) Human Dobrava hantavirus infections in Estonia. Lancet 352:369
Lundkvist Å, Wiger D, Hörling J, Brus Sjölander K, Plyusnina A, Mehl R, Vaheri A, Plyusnin A (1998b) Isolation and characterization of Puumala hantavirus from Norway: evidence for a distinct phylogenetic sublineage. J Gen Virol 78:2603–2614
Mäkelä S, Hurme M, Ala-Houhala I, Mustonen J, Partanen J, Vapalahti O, Vaheri A, Pasternack A (2000) Polymorphism of the cytokine genes in hospitalized patients with Puumala hantavirus infection. Submitted (Nephrol Dial Transpl)
Malkovsky M (1996) HLA and natural history of HIV infection. Lancet 348:142–143
McBride BW, Ward KA (1987) Herpes simplex-specific IgG subclass response in herpetic keratitis. J Med Virol 21:179–189
McGuire W, Hill AVS, Allsopp CEM, Greenwood BM, Kwjatkowski D (1994) Variation in the TNF-α promoter region associated with susceptibility to cerebral malaria. Nature 371:508–511
Meisel H, Lundkvist Å, Gantzer K, Bär W, Sibold C, Krüger DH (1998) First clinical case of infection by Dobrava hantavirus in Germany. Eur J Clin Microbiol Infect Dis 17:884–885
Mertz GJ, Hjelle BL, Williams TM, Koster FT (1999) Host responses in the hantavirus cardiopulmonary syndrome. In: Saluzzo JF, Dodet B (eds) Factors in the emergence and control of rodent-borne diseases. Elsevier, Paris, pp 133–137
Morgan BP, Walport MJ (1991) Complement deficiency and disease. Immunol Today 12:301–306
Mori M, Rothman AL, Kurane I, Montoya JM, Nolte KB, Norman JE, Waite DC, Koster FT, Ennis FA (1999) High levels of cytokine-producing cells in the lung tissues of patients with fatal hantavirus pulmonary syndrome. J Infect Dis 179:295–302
Morii M, Yoshimatsu K, Arikawa J, Zhou G, Kariwa H, Takashima I (1998) Antigenic characterization of Hantaan and Seoul virus nucleocapsid proteins expressed by recombinant baculovirus: application

of a truncated protein, lacking an antigenic region common to the two viruses, as a serotyping antigen. J Clin Microbiol 36:2514–2521
Mustonen J, Partanen J, Kanerva M, Pietilä K, Vapalahti O, Pasternack A, Vaheri A (1996) Genetic susceptibility to severe course of nephropathia epidemica caused by Puumala hantavirus. Kidney Int 49:217–221
Mustonen J, Partanen J, Kanerva M, Pietilä K, Vapalahti O, Pasternack A, Vaheri A (1998) Asssociation of HLA B27 with benign clinical course of nephropathia epidemica caused by Puumala hantavirus. Scand J Immunol 47:277–279
Nadel S, Newport MJ, Booy R, Levin M (1996) Variation in the tumor necrosis factor-alpha gene promoter region may be associated with death from meningococcal disease. J Infect Dis 174:878–880
Nemirov K, Vapalahti O, Lundkvist Å, Vasilenko V, Golovljova I, Plyusnina A, Niemimaa J, Laakkonen J, Henttonen H, Vaheri A, Plyusnin A (1999) Isolation and characterization of Dobrava hantavirus carried by the striped field mouse (*Apodemus agrarius*) in Estonia. J Gen Virol 80:371–379
Nilsen I, Flaegstad T, Traavik T (1991) Detection of specific IgA antibodies against BK virus by ELISA. J Med Virol 33:89–94
Nolte KB, Feddersen RM, Foucar K, Zaki SR, Koster FT, Madar D, Merlin TL, McFeeley PJ, Umland ET, Zumwalt RE (1995) Hantavirus pulmonary syndrome in the United States: a pathological description of a disease caused by a new agent. Hum Pathol 26:110–120
Padula PJ, Rossi CM, Della Valle MO, Martinez PV, Colavecchia SB, Edelstein A, Miguel SD, Rabinovich RD, Segura ELJ (2000) Development and evaluation of a solid-phase enzyme immunoassay based on Andes hantavirus recombinant nucleoprotein. Med Microbiol 49:149–155
Plyusnin A, Hörling J, Kanerva M, Mustonen J, Cheng Y, Partanen J, Vapalahti O, Kukkonen SKJ, Niemimaa J, Henttonen H, Niklasson B, Lundkvist Å, Vaheri A (1997) Puumala hantavirus genome in patients with nephropathia epidemica: correlation of PCR positivity with HLA haplotype and link to viral sequences in local rodents. J Clin Microbiol 35:1090–1096
Plyusnin A, Mustonen J, Asikainen K, Plyusnina A, Niemimaa J, Henttonen H, Vaheri A (1999) Analysis of Puumala hantavirus genome in patients with nephropathia epidemica and rodent carriers from the sites of infection. J Med Virol 59:397–405
Roy S, McGuire W, Mascie-Taylor CG, Saha B, Hazra SK, Hill AV, Kwiatkowski D (1997) Tumor necrosis factor promoter polymorphism and susceptibility to lepromatous leprosy. J Infect Dis 176: 530–532
Sällberg M, Norder H, Magnius LO (1990) Comparison of class and subclass distribution of antibodies to the hepatitis B core and B e antigens in chronic hepatitis B. J Med Virol 30:1–6
Schmaljohn CS, Chu YK, Schmaljohn AL, Dalrymple JM (1990) Antigenic subunits of Hantaan virus expressed by baculovirus and vaccinia virus recombinants. J Virol 64:3162–3170
Settergren B, Ahlm C, Juto P, Niklasson B (1991) Specific Puumala virus IgG half a century after haemorrhagic fever with renal syndrome. Lancet 338:66
Sibold C, Meisel H, Lundkvist Å, Schulz A, Cifire F, Ulrich R, Kozuch O, Labuda M, Krüger DH (1999) Simultaneous occurrence of Dobrava, Puumala and Tula hantaviruses in Slovakia, Central Europe. Am J Trop Med Hyg 61:409–411
Sidelnikov YN, Sivoraksha GA (1990) The participation of histamine and serotonin in the genesis of acute kidney failure in patients with hemorrhagic fever with renal syndrome. Urol Nefrol Mosk 4: 46–48
Steel CM, Ludlam CA, Beatson D, Peutherer JF, Cuthbert RJG, Simmonds P, Morrison H, Jones M (1988) HLA haplotype A1 B8 DR3 as a risk factor for HIV-related disease. Lancet 1:1185–1188
Takala A, Lähdevirta J, Jansson S-E, Vapalahti O, Orpana A, Karonen S-L, Repo H (2000) Systemic inflammation in hemorrhagic fever with renal syndrome correlates with hypotension and thrombocytopenia but not with renal injury. J Infect Dis 181:1964–1970
Temonen M, Mustonen J, Helin H, Pasternack A, Vaheri A, Holthöfer H (1996) Cytokines, adhesion molecules, and cellular infiltration in nephropathia epidemica kidneys: An immunohistochemical study. Clin Immunol Immunopathol 78:47–55
Tiwari JL, Terasaki PI (1985) HLA and disease associations. Springer-Verlag, New York
Ulrich R, Lundkvist Å, Meisel H, Koletzki D, Brus Sjölander K, Gelderblom HR, Borisova G, Schnitzler P, Darai G, Krüger DH (1998) Chimeric HBV core particles carrying a defined segment of Puumala hantavirus nucleocapsid protein evoke protective immunity in an animal model. Vaccine 16:272–280
Van Epps HL, Schmaljohn CS, Ennis FA (1999) Human memory cytotoxic T-lymphocyte (CTL) responses to Hantaan virus infection: identification of virus-specific and cross-reactive CD8(+) CTL epitopes on nucleocapsid protein. J Virol 73:5301–5308

Vapalahti O, Kallio-Kokko H, Närvänen A, Julkunen I, Lundkvist Å, Plyusnin A, Lehväslaiho H, Brummer-Korvenkontio M, Vaheri A, Lankinen H (1995) Human B-cell epitopes of Puumala virus nucleocapsid protein, the major antigen in early serological response. J Med Virol 46:293–303

Vapalahti O, Lundkvist Å, Kallio-Kokko H, Paukku K, Julkunen I, Lankinen H, Vaheri A (1996) Antigenic properties and diagnostic potential of Puumala virus nucleocapsid protein expressed in insect cells. J Clin Microbiol 34:119–125

Vapalahti O, Lundkvist Å, Fedorov V, Conroy CJ, Hirvonen S, Plyusnina A, Nemirov K, Fredga K, Cook JA, Niemimaa J, Kaikusalo A, Henttonen H, Vaheri A, Plyusnin A (1999) Isolation and characterization of a hantavirus from *Lemmus sibiricus*: evidence for host switch during hantavirus evolution. J Virol 73:5586–5592

Wang M, Pennock DG, Spik KW, Schmaljohn CS (1993) Epitope mapping studies with neutralizing and non-neutralizing antibodies to the G1 and G2 envelope glycoproteins of Hantaan virus. Virology 197:757–766

Wilson AG, Symons JA, McDowell TL, McDevit HO, Duff GW (1997) Effects of a polymorphism in the human tumor necrosis factor alpha promoter on transcriptional activation. Proc Natl Acad Sci 94:3195–3199

Yang CW, Bang BK (1992) Changes of tumor necrosis factor-alpha in patients with hemorrhagic fever with renal syndrome. J Catholic Med College 45:819–830

Yoshimatsu K, Yoo YC, Yoshida R, Ishihara C, Azuma I, Arikawa J (1993) Protective immunity of Hantaan virus nucleocapsid and envelope protein studied using baculovirus expressed proteins. Arch Virol 130:365–376

Yoshimatsu K, Arikawa J, Tamura M, Yoshida R, Lundkvist Å, Niklasson B, Kariwa H, Azuma I (1996) Characterization of the nucleocapsid protein of Hantaan virus strain 76–118 using monoclonal antibodies. J Gen Virol 77:687–694

Zaki SR, Greer PW, Coffield LM, Goldsmith CS, Nolte KB, Foucar K, Feddersen RM, Zumwalt RE, Miller GL, Khan AS, et al. (1995) Hantavirus pulmonary syndrome. Pathogenesis of an emerging infectious disease. Am J Pathol 146:552–579

Zhang XK, Takashima I, Mori F, Hashimoto N (1989) Comparison of virulence between two strains of Rattus serotype hemorrhagic fever with renal syndrome (HFRS) virus in newborn rats. Microbiol Immunol 33:195–205

Zöller L, Yang S, Gött P, Bautz EKF, Darai G (1993) A novel μ-capture EIA based on recombinant proteins for sensitive and specific diagnosis of hemorrhagic fever with renal syndrome. J Clin Microbiol 31:1194–1199

Vaccines Against Hantaviruses

J.W. Hooper[1] and D. Li[2]

1 Introduction

Hantaviruses are associated with two human diseases: hemorrhagic fever with renal syndrome (HFRS) and hantavirus pulmonary syndrome (HPS). These viruses are carried by persistently infected rodents, and are transmitted to humans by aerosolized rodent excreta. HFRS is characterized by fever, hemorrhagic manifestations, renal dysfunction, and shock. HPS is characterized by fever, headache, myalgia, and gastrointestinal symptoms, followed by noncardiogenic pulmonary edema, and shock. Although there are orders of magnitude fewer HPS cases than HFRS cases, the small outbreaks of HPS that have occurred have been highly lethal, with case-fatality rates of 40–60%. In addition, there is evidence that at least

[1] Virology Division, United States Army Medical Research Institute for Infectious Diseases, Fort Detrick, Maryland 21702, USA
[2] Institute of Virology, Chinese Academy of Preventive Medicine, 100 Yiny Xin Jie, Xuan Wu Qu. Beijing 100052, China

one of the HPS hantaviruses can be transmitted from person-to-person (Wells et al. 1997). Since 1993, 555 HPS cases have been reported in the United States, Canada, and several countries in South America. Recently, for the first time, a hantavirus was associated with an outbreak of lethal acute febrile respiratory illness in Central America (Bayard et al. 2000). A summary of hantaviruses that are known to cause human disease are shown in Table 1.

Hantaviruses are members of the family *Bunyaviridae*, genus *Hantavirus*. The viral genome consists of three segments of negative-strand RNA. The large (L) segment encodes the RNA-dependent RNA polymerase; the medium (M) segment encodes the envelope glycoproteins (G1, and G2); and the small (S) segment encodes the nucleocapsid protein (N). The genome segments are complexed with N to form ribonucleoprotein (RNP) structures that are packaged within the G1/G2-containing viral envelope. G1/G2 play important roles in the biological functions of hantaviruses, including attachment to receptors on cells and membrane fusion events; they are also the targets of neutralizing antibodies (Schmaljohn 1996).

The role(s) that L, N, and G1/G2 play in eliciting protective immunity remain under of investigation. Neutralizing antibodies to G1 and G2, but not N, have been identified. Several G1- or G2-specific monoclonal antibodies confer protection in passive transfer experiments in rodents (Schmaljohn et al. 1990; Arikawa et al. 1992). Although N does not elicit high-titer neutralizing antibodies, several studies, described below, indicate that N does play a role in eliciting protective immunity.

Table 1. Hantaviruses associated with human disease

Disease	Virus	Abbreviation	Distribution	Number of cases	Case-fatality rate
HFRS				50–150,000 per year	0.1–10%
	Hantaan virus	HTNV	Asia, Far East Russia		
	Seoul virus	SEOV	Asia[a]		
	Dobrava virus	DOBV	Balkans, Europe		
	Puumala virus	PUUV[b]	Eurasia[b]		
HPS				~555 cases since 1993	40–60%
	Sin nombre virus	SNV	United States		
	Andes virus	ANDV	Argentina, Chile		
	Black creek canal virus	BCCV	Florida		
	Bayou virus	BAYV	South Western U.S.		
	Laguna Negra virus	LNV	Paraguay, Bolivia		
	Juquitiba	JUQV	Brazil		
	Araraquara	ARAV	Brazil		
	Castelo dos Sonhos	CASV	Brazil		
	Choclo (tentative)	CHOV	Panama		

[a] Most cases are in Asia, but SEOV is found worldwide;
[b] PUUV causes a mild form of HFRS known as nephropathia epidemica with a case-fatality rate of 0.1–0.3%.

There is no evidence that L elicits an immune response during natural infection; however, this possibility has not been ruled out.

2 Recombinant DNA Approaches to Vaccine Development

Hantaan virus (HTNV) was identified as the etiological agent causing HFRS in 1978 at approximately the same time that technological advances in molecular biology allowed researchers to clone, manipulate, and express individual genes and gene fragments for the first time. The advent of recombinant DNA technology has allowed researchers to approach vaccine development in ways that differ from those of the more traditional killed or attenuated virus vaccines. Hantavirus researchers are investigating several different recombinant DNA strategies to develop novel vaccines that protect against one or more hantaviruses. These approaches include recombinant baculovirus-expressed proteins, *Escherichia coli*-expressed proteins, chimeric and mosaic hepatitis B core antigens, recombinant vaccinia viruses, packaged alphavirus replicons, and naked DNA. Investigators have evaluated the immunogenicity and, in some cases, the protective efficacy of individual hantavirus proteins or combinations of proteins. Table 2 summarizes recombinant DNA vaccines that have been tested for protective efficacy in rodent infection models. Rodent infection models have been used to assess protective efficacy of hantavirus vaccines because there is currently no practical disease model in adult animals for either HFRS or HPS.

2.1 Recombinant Baculovirus-Infected Cell Lysate

The first studies that investigated the immunogenicity of individual hantavirus proteins were performed using partially purified extracts derived from insect cells infected with recombinant baculoviruses. Mice vaccinated with baculovirus-derived HTNV N or G1/G2 developed antibody responses as measured by the indirect fluorescent antibody test (IFAT); however, neutralizing antibody titers were low or undetectable (Schmaljohn et al. 1990; Yoshimatsu et al. 1993). Hamsters vaccinated with baculovirus-derived N, G1, G2, or G1/G2 all developed antibodies detectable by either enzyme-linked immunosorbent assay (ELISA) or radioimmunoprecipitation assay (RIPA), but only the G1/G2-vaccinated animals developed neutralizing antibodies as measured by the plaque reduction-neutralization test (PRNT; Schmaljohn et al. 1990) (Table 3). Bank voles vaccinated with baculovirus-derived PUUV N also produced an antibody response as detected by IFAT (Lundkvist et al. 1996).

The protective efficacy of the recombinant baculovirus antigens were tested by vaccinating rodents with baculovirus-infected cell lysate, challenging the

Table 2. Recombinant DNA vaccines against hantaviruses: protecting rodents against homotypic hantaviruses infection

Vaccine	Route[a] and schedule	Adjuvant	Vaccinee species	Challenge virus/route[a]	Protection[b] (number of protected animals/total tested)						Reference
					Immunogen						
					N	G1	G2	G1/G2	N & G1/G2 mix	N/G1/G2	
Recombinant baculovirus cell lysate	I.m., 2 vacc at 28-day interval	–	Hamster	HTNV/i.m.	YES (5/5)[c]	MOST (3/5)[c]	SOME (1/5)[c]	YES (5/5)[c]			Schmaljohn et al. 1990
	S.c., 3 vacc at 3-week intervals	+	Bank vole	PUUV/s.c.	YES (8/8)[d]						Lundkvist et al. 1996
Purified protein expressed in *E. coli*	S.c., 3 vacc at 3-week intervals	+	Bank vole	PUUV/s.c.	MOST (4/5)[d] aa 1–79 YES (3/3) aa 1–118 YES (3/3) aa 1–267 MOST (2/3) aa 229–327						Lundkvist et al. 1996
Purified chimeric HBV VLP	S.c., 3 vacc at 3-week intervals	+	Bank vole	PUUV/s.c.	MOST (4/5)[e] aa 1–45 NO (0/5) aa 38–82 SOME (1/4) aa 75–119						Ulrich et al. 1998
	S.c., 3 vacc at 3-week intervals	–	Bank vole	PUUV/s.c.	SOME (2/4)[f] aa 1–45 NO (0/3) aa 38–82 NO (0/3) aa 38–82						Ulrich et al. 1999
Purified mosaic HBV VLP	S.c., 3 vacc at 3-week intervals	– +	Bank vole	PUUV/s.c.	SOME (1/4)[5] aa 1–114 NO (0/2) aa 1–114						Ulrich et al. 1999

Recombinant vaccinia virus infection	Scarification, one i.p. boost on day 28	Hamster	HTNV/i.m.	SOME (2/5)[g]	NO (0/5)[g]	MOST (3/5)[g]		YES (9/9)[g]		Schmaljohn et al. 1990
	I.m., 2 vacc at 30-day interval	Hamster	HTNV/i.m.						YES (13/13)[h]	Chu et al. 1995
	S.c., 2 vacc at 28-day interval	Gerbil	SEOV/s.c.	MOST (3/4)[i]			YES (4/4)[i]	YES (4/4)[i]		Xu et al. 1992
Packaged Sindbis replicon	S.c., 3 vacc at 4-week intervals	Hamster	SEOV/i.m.	NO (0/4)[j]			SOME (1/5)[k]			Kamrud et al. 1999
DNA vaccination	Gene gun, 3 vacc at 4-week intervals	Hamster	SEOV/i.m.	NO (0/5)[j]			YES (5/5)[k]			Hooper et al. 1999
	Gene gun, 3 vacc at 3-week intervals	Hamster	SEOV/i.m.	SOME (1/9)[j]			YES (6/6)[k]			Kamrud et al. 1999
DNA-launched Sinbis replicon	Gene gun, 3 vacc at 3-week intervals	Hamster	SEOV/i.m.	SOME (2/8)[j]			MOST (3/5)[k]			Kamrud et al. 1999

[a] Abbreviations: vacc, vaccination; i.m., intramuscular needle injection; s.c., subcutaneous needle injection; i.p., intraperitoneal;
[b] YES, complete protection; MOST, > 50% protection; SOME, > 0% protection; NO, 0% protection;
[c] Protected/total, as measured by antigen in the lungs, or G1/G2-specific antibody detected by RIPA;
[d] Protected/total, as measured by antigen in the lungs, or G1/G2-specific antibody as measured by capture ELISA;
[e] Protected/total, as measured by antigen in the lungs or G2-specific antibody detected by capture ELISA, or FRNT;
[f] Protected/total, as measured by a G2-specific antibody ELISA;
[g] Protected/total, as measured by antigen in the lungs, or N-specific antibody detected by RIPA;
[h] Protected/total, as measured by antigen in the lungs or kidneys, or hantavirus RNA in the lungs or kidneys (RT-nested PCR);
[i] Protected/total, as measured by virus isolation from the lungs;
[j] Protected/total, as measured by G1/G2-specific IFAT or PRNT;
[k] Protected/total, as measured by postchallenge N-specific antibody by ELISA.

Table 3. Recombinant DNA vaccines eliciting hantavirus neutralizing antibody responses

Vaccine	Route[a] and schedule	Vaccinee species	Immunogen source	Immunogen	Neutralizing antibody positive/total vaccinees (titer)[b]	Reference
Recombinant baculovirus cell lysate	I.m., 2 vacc at 28-day interval	Hamster	HTNV	G1 G2 G1/G2	0/5 0/5 5/5	Schmaljohn et al. 1990
Recombinant vaccinia virus infection	Scarification, one i.p. boost on day 14	BALB/c mice	HTNV	G1/G2	5/5 ($MT_{80\%}=320$)	Penserio et al. 1988
	Scarification, one i.p. boost on day 28	Hamster	HTNV	G1 G2 G1/G2	0/5 0/5 5/5	Schmaljohn et al. 1990
	I.m., 2 vacc at 30-day interval	Hamster	HTNV	G1/G2/N	50/50	Chu et al. 1995
	S.c., 2 vacc at 28-day interval	Gerbil	SEOV	G1/G2	4/4 ($MT_{50\%}=200$)	Xu et al. 1992
				G1/G2 & N mix	4/4 ($MT_{50\%}=200$)	
	S.c., one boost on week 6	Human	HTNV	G1/G2/N	31/43 ($GMT_{50\%}=160$)	McClain et al. 2000
Packaged Sindbis replicon	S.c., two boosts at 4-week intervals	Hamster	SEOV	G1/G2	4/5 ($GMT_{80\%}=127$)	Kamrud et al. 1999
DNA vaccination	Gene gun, 3 vacc at 4-week intervals	BALB/c mice	SEOV	G1/G2	10/10 ($GMT_{50\%}=260$)	Hooper et al. 1999
	Gene gun, 3 vacc at 4-week intervals	Hamster	SEOV	G1/G2	5/5 ($GMT_{80\%}=366$)	Hooper et al. 1999
	Gene gun, 3 vacc at 3-week intervals	Hamster	SEOV	G1/G2	6/6 ($GMT_{80\%}=3225$)	Kamrud et al. 1999
	I.m., 3 vacc at 4-week intervals	BALB/c mice	SEOV	G1/G2	1/10[c]	Hooper et al. 1999
	I.m., 3 vacc at 4-week intervals	BALB/c mice	SNV	Fragments of G1	0/5–5/5[d]	Bharadwaj et al. 1999
				Fragments of G2	0/5–5/5[e]	
DNA-launched Sinbis replicon	Gene gun, 3 vacc at 3-week intervals	Hamster	SEOV	G1/G2	5/5 ($GMT_{80\%}=211$)	Kamrud et al. 1999

Table 3. (*Contd.*)

[a] Abbreviations: vacc, vaccination; i.m., intramuscular needle injection; s.c., subcutaneous needle injection; i.p., intraperitoneal;
[b] Titer as mean titer (MT) or geometric mean titer (GMT). Titer based on the lowest dilution neutralizing 50% or 80% of plaques or foci;
[c] Only 1 of 10 mice gave a PRNT50% of greater than 1:40. Mouse serum diluted less than 1:40 gave inconsistent, irreproducible results;
[d] Sera from mice vaccinated with G1 aa 1–167, 265–432, 399–565, 486–652, but not 132–299 exhibited FRNT50% titers when diluted 1:10;
[e] Sera from mice vaccinated with G2 aa 1–167 and 134–300, but not 266–433 or 400–488 exhibited FRNT50% titers when diluted 1:10.

animals, and then monitoring them for evidence of hantavirus infection. Hamsters vaccinated with either HTNV N, G1, G2, or G1/G2 were uniformly protected against infection with HTNV as measured by the absence of viral antigen in the lungs; however, G1 and G2 alone did not completely protect the hamsters as measured by postchallenge detection of N-specific antibodies (SCHMALJOHN et al. 1990). Bank voles vaccinated with PUUV N were uniformly protected against infection with PUUV as measured by the absence of viral antigen in the lungs and the absence of an antibody response to G1/G2 as measured by ELISA (LUNDKVIST et al. 1996).

2.2 *E. coli*-Expressed Truncated Nucleocapsid as Immunogen

To determine which region(s) of the exogenously expressed N protein was immunogenic and protective, Lundkvist et al. vaccinated bank voles with truncated versions of the PUUV N expressed in *E. coli* (LUNDKVIST et al. 1996). Truncated N immunogens (aa 1–79, 1–118, 1–267, and 229–327) were all immunogenic, and all conferred either complete (aa 1–118, 1–267) or partial (aa 1–79, 229–327) protection against infection in bank voles (LUNDKVIST et al. 1996). In similar experiments, ULRICH et al. demonstrated that truncated versions of the Puumala N, expressed as antigens within chimeric hepatitis B virus core antigen (HBVcAg) virus-like particles, were all immunogenic, and either protected bank voles partially (aa 1–45, 75–119), or failed to protect them (aa 38–82; ULRICH et al. 1998). The HBVcAg virus-like particles consist of *E. coli*-expressed, truncated HBV core antigen containing the aforementioned PUUV N sequences. In the absence of adjuvant, the protective effect of HBVcAg virus-like particles containing aa 1–45 was greatly diminished (ULRICH et al. 1999). Mosaic HBVcAg virus-like particles, in which particles contain both HBVcAg and HBVcAg-fusion proteins, were capable of expressing more PUUV N epitopes (larger insertion capacity); however, the protective efficacy of the mosaic particles was reduced relative to chimeric particles (ULRICH et al. 1999).

The mechanism by which vaccination with exogenously expressed N protects against infection remains unknown. Although vaccination with N elicits antibody responses, in most cases these antibodies fail to efficiently neutralize virus in vitro. There is one report that vaccination with certain fragments of *E. coli*-expressed PUUV N elicited low-titer neutralizing antibody in bank voles, as measured by focus reduction assay (LUNDKIVST et al. 1996). Protection in the absence of high-titer anti-N neutralizing antibody suggests that a protective cell-mediated (i.e., cytotoxic T cell) immune response could be elicited by vaccination with exogenously expressed N.

2.3 Recombinant Vaccinia Virus Hantavirus Vaccines

Recombinant vaccinia viruses have been tested as vaccine platforms for many other pathogens including hantaviruses. Recombinant vaccinia viruses, containing the gene(s) of interest controlled by a vaccinia promoter(s), are administered to the vaccinee as infectious virus. The recombinant viruses infect host cells and express the protein of interest, which is then presented to the immune system in a context that may be significantly different from the presentation of exogenously expressed protein (e.g., recombinant baculovirus-infected cell lysate, *E. coli*-expressed protein, or HBV chimeric antigen).

Recombinant vaccinia viruses expressing the HTNV N, or G1/G2 were tested for immunogenicity in mice (PENSIERO et al. 1988). An antibody response was detected in all cases, but only the G1/G2-expressing recombinant elicited neutralizing antibodies (Table 3). When recombinant vaccinia viruses expressing G1 alone or G2 alone were tested in mice, antibody responses were detected by ELISA and RIPA; however, PRNT results were not reported (SCHMALJOHN et al. 1990).

Vaccinia recombinants expressing HTNV N, G1, G2, or G1/G2 were also tested for immunogenicity in hamsters and gerbils. Recombinants expressing N, G1, G2, or G1/G2 elicited antibody responses in hamsters as measured by ELISA and RIPA; however, as in the case of recombinant baculoviruses, only the G1/G2 elicited neutralizing antibody (SCHMALJOHN et al. 1990). Similarly, Seoul virus (SEOV) N or G1/G2 vaccinia recombinants elicited antibody responses in gerbils as measured by ELISA, but only G1/G2 elicited a neutralizing antibody response in these animals (XU et al. 1992) (Table 3).

The protective efficacy of recombinant vaccinia viruses expressing HTNV or SEOV immunogens were tested in hamsters or gerbils, respectively (Table 2). The vaccinia recombinants expressing the HTNV G1/G2 completely protected the animals, and vaccinating them with G1, G2, or N conferred partial protection (SCHMALJOHN et al. 1990). Vaccinia recombinants expressing the SEOV G1/G2 or co-vaccination with G1/G2 and N expressing viruses, conferred complete protection, and N alone partially protected gerbils against SEOV infection (XU et al. 1992).

The recombinant vaccinia vaccine expressing both Hantaan N and G1/G2 (SCHMALJOHN et al. 1992) was tested in phase I and phase II clinical safety studies (McCLAIN et al. 2000). This was the first hantavirus recombinant DNA vaccine tested in humans. The vaccine was safe and caused only local reactogenicity. Immunogenicity was greatly influenced by preexisting immunity to vaccinia virus (i.e., vaccination with the smallpox vaccine). In the phase I dose escalation study, among volunteers receiving subcutaneous inoculation of the full-strength dose of the vaccine (3.4×10^7pfu) 44% (7 of 16) of the vaccinia-naïve vaccinees exhibited a PRNT 50% titer $\geqslant 10$ after one vaccination, and 100% (6 of 6) after two vaccinations (42-day interval). In contrast, only 33% (2 of 6) vaccinia preimmune vaccinees exhibited neutralizing antibody responses after one vaccination, and 50% (3 of 6) had detectable titers after two vaccinations.

In the phase II expanded safety trial, 72% of 43 vaccinia-naive volunteers developed neutralizing antibodies to HTNV after two vaccinations (GMT = 160); however, only 26% of the 47 vaccinia-immune volunteers showed neutralizing antibody responses (GMT = 24) to HTNV (McCLAIN et al. 2000). ELISAs were performed to detect antibody responses not detected by the PRNT (e.g., anti-N, or non-neutralizing anti-G1/G2 responses). Five PRNT-negative vaccinees were positive by ELISA. Together with the PRNT results, these data indicated that 84% of vaccinia-naïve individuals, versus 32% of the vaccinia-preimmune individuals, seroconverted after two vaccinations. Lymphocyte proliferation assays were performed with peripheral blood lymphocytes (PBLC) from individual vaccinees. The results appeared to correlate with antibody responses: vaccinia-naïve individuals exhibited a higher stimulation index than vaccinia-preimmune individuals. Whether N, G1/G2, or both immunogens elicited the lymphoproliferative response was not determined. Interestingly, the stimulation index was elevated not only when Hantaan antigen was used to stimulate the PBLC, but also when Sin Nombre virus (SNV) was used. These data suggest that the cell-mediated response elicited by the Hantaan vaccine might cross-react with other hantaviruses.

2.4 Packaged Sindbis Virus Replicons

Although vaccinia recombinant vaccines show promise as vaccines against hantaviruses, especially in vaccinia-naïve individuals, there are drawbacks to vaccination with live viruses, which include the development of lesions containing infectious virus, scaring, and potentially life-threatening disseminated vaccinia infection in immunocompromised individuals. Packaged alphavirus replicons provide a way to deliver genes encoding immunogens of interest into host cells (i.e., mimic infection) without using replication-competent virus as a vector (FROLOV et al. 1996). Alphavirus replicons consist of a self-replicating viral RNA that contains a gene of interest controlled by an alphavirus promoter. Although the RNA is self-replicating, it is not infectious because the viral genes that encode structural proteins have been removed. Replicon RNA is packaged in alphavirus structural proteins supplied by helper RNA.

Packaged Sindbis virus replicons expressing the SEOV N or G1/G2 proteins were generated, purified, and used to vaccinate hamsters. Hamsters vaccinated with N replicons developed an N-specific non-neutralizing antibody response as measured by ELISA, but were not protected against infection with SEOV because the animals developed G1/G2-specific antibodies as measured by IFAT and PRNT, after challenge (Kamrud et al. 1999). Most of the hamsters vaccinated with G1/G2 replicons developed neutralizing antibody responses, and most were protected from infection as measured by the absence of an N-specific antibody response after challenge (Table 2).

2.5 Naked DNA

Another way to avoid the inherent problems associated with vaccination with a vaccinia recombinant virus, and yet still accomplish intracellular expression of the hantavirus immunogens, is to deliver the vaccine as naked DNA. This technique, known as DNA vaccination or genetic vaccination, involves delivering plasmid DNA, containing a gene of interest controlled by a promoter that directs expression within mammalian cells, into the cells of a vaccinee. A common promoter used in this technology is the cytomegalovirus immediate-early (CMV IE) promoter. The expressed protein/immunogen is presented to the immune system and elicits an immune response. This technology has been used to confer protective immunity to numerous pathogens (Robinson et al. 1997).

Hooper et al. demonstrated that SEOV M or S genome segment naked DNA constructs, expressing G1/G2 or N proteins, respectively, were immunogenic in mice (Hooper et al. 1999). The DNA vaccines were delivered into the muscle by needle injection (intramuscular inoculation, i.m.), or into the epidermis by a gene gun. Although both i.m. and gene gun administration elicited antibody responses, the gene gun route elicited a more consistent, higher-titer response. For instance, a neutralizing antibody response was detected in 10 of 10 mice vaccinated with M by gene gun, but only 1 of 10 mice vaccinated by i.m. needle injection. The gene gun route was used in subsequent protection studies in hamsters.

The SEOV M and S naked DNA gene gun vaccines were tested for immunogenicity and protective efficacy in hamsters (Hooper et al. 1999). Both the M and S DNA vaccines were immunogenic. Vaccination with the M gene (G1/G2 immunogens) elicited high-titer neutralizing antibody (GMT = 366), and vaccination with the S gene (N immunogen) elicited an antibody response as measured by an N-specific ELISA, or IFAT. Although both vaccines were immunogenic, only the M gene vaccine (G1/G2) protected the hamsters from infection, as measured by the absence of a postchallenge anti-N antibody response. The S gene vaccine (N) failed to protect against infection because the animals exhibited high-titer neutralizing antibody responses postchallenge.

These data demonstrated that, unlike exogenously expressed N, naked DNA-expressed N (endogenously expressed) failed to confer protection against infection. However, it must be noted that the absence of "sterile" protection does not rule out the possibility that the anti-N immune response could limit the infection, clear the infection, or prevent disease.

Bharadwaj et al. reported that fragments of SNV G1 or G2 elicited both humoral and cell-mediated immunity (BHARADWAJ et al. 1999). BALB/c mice vaccinated i.m. with naked DNA expressing overlapping fragments (~167 amino acids) of G1/G2 developed low levels (⩽1:20 to give a 50% reduction in plaque number) of neutralizing antibody as measured by focus reduction neutralization test (FRNT). This is an interesting finding because, in all previous reports, neutralizing antibody responses were only elicited when both G1 and G2 were included in the vaccine. In addition to a neutralizing antibody response, fragments of G1/G2 also elicited a cell-mediated immune response as measured by splenocyte proliferation upon stimulation with cognate antigens expressed in *E. coli*. Fragments that elicited the greatest level of proliferation were not necessarily the same fragments that elicited neutralizing antibody. A naked DNA vaccine expressing the full-length SNV N protein elicited an antibody response and proliferative response in BALB/c mice (BHARADWAJ et al. 1999). The protective efficacies of the aforementioned SNV naked DNA constructs have not yet been reported.

2.6 DNA-Launched Sindbis Virus Replicons

DNA-launched alphavirus replicons are essentially naked DNA plasmids that transcribe a self-replicating alphavirus RNA containing a gene of interest. Several alphavirus replicons have been engineered into DNA-based expression vectors, including a Sindbis virus-based construct (DUBENSKY et al. 1996). Sindbis replicons containing the SEOV S or M gene were constructed, tested for expression in cell culture, and then tested for immunogenicity and protective efficacy in hamsters (KAMRUD et al. 1999). Both the S and M constructs were immunogenic in hamsters, as measured by anti-N ELISA or PRNT, respectively. Two of eight of the S vaccinated hamsters were protected and three of five of the M vaccinated hamsters were protected (Table 2). In a comparative study, the DNA-launched Sindbis virus replicon expressing SEOV N elicited higher ELISA titers (GMT = 6400, $n = 8$) than naked DNA (GMT = 864, $n = 9$), or packaged Sindbis replicons (GMT = 1600, $n = 4$) However, the DNA-launched Sindbis replicon expressing SEOV M elicited lower PRNT titers (GMT = 211, $n = 5$) than naked DNA (GMT = 3225, $n = 6$), or packaged Sindbis replicons (GMT = 1270, $n = 5$) (Table 3) (KAMRUD et al. 1999). It remains to be determined if DNA-launched alphavirus replicons will confer significant advantages over conventional naked DNA vaccines.

2.7 Cross-Protection with Recombinant DNA Vaccines

Recombinant vaccinia viruses expressing SEOV G1/G2, or N, or co-infection with both recombinants cross-protected gerbils against HTNV (XU et al. 1992; Table 4). Recombinant vaccinia virus expressing HTNV N and G1/G2 cross-protected

Table 4. Recombinant DNA vaccines against hantaviruses: cross-protection against heterotypic hantaviruses

Vaccine	Route[b] and schedule	Vaccine species	Source of Immunogen	Challenge virus/route[b]	Protection[a] (number of protected animals/total tested)				Reference
					Immunogen				
					N	G1/G2	N & G1/G2 mix	N/G1/G2	
Recombinant vaccinia virus infection	I.m., 2 vacc at 30-day interval	Hamster	HTNV	SEOV/i.m.				YES (13/13)[c]	Chu et al. 1995
			HTNV	PUUV/i.m.				SOME (2/13)[c]	
	S.c. injection with one boost on day 28	Gerbil	SEOV	HTNV/i.m.	SOME (1/3)[d]	YES (4/4)[d]	YES (3/4)[d]		Xu et al. 1992
DNA vaccination	Gene gun, 3 vacc at 3-week intervals	Hamster	SEOV	HTNV/i.m.		MOST (4/5)[e]			Kamrud et al. 1999
	Gene gun, 3 vacc at 3-week intervals	Hamster	SEOV	HTNV/i.m.		MOST (4/5)[e]			J.W. Hooper, unpublished data
	Gene gun, 3 vacc at 3-week intervals	Hamster	SEOV	DOBV/i.m.		MOST (7/10)[e]			J.W. Hooper, unpublished data
	Gene gun, 3 vacc at 3-week intervals	Hamster	SEOV	PUUV/i.m.		SOME (1/5)[e]			J.W. Hooper, unpublished data

[a] YES, complete protection; MOST, > 50% protection; SOME, > 0% but < 50% protection; NO, 0% protection;
[b] Abbreviations: vacc, vaccination; i.m., intramuscular needle injection;
[c] As measured by antigen in the lungs or kidneys, or hantavirus RNA in the lungs or kidneys (RT-nested PCR);
[d] As measured by virus isolation from the lungs;
[e] As measured by postchallenge N-specific antibody (ELISA).

hamsters against SEOV but not against Puumala virus (PUUV) (Chu et al. 1995). Similarly, naked DNA containing the SEOV M gene (expressing G1/G2) cross-protected hamsters against HTNV (Kamrud et al. 1999; J.W. Hooper, unpublished data), and Dobrava virus (DOBV) but not against PUUV (J.W. Hooper, unpublished data). In naked DNA cross-protection experiments, SEOV M-vaccinated hamsters with anti-SEOV $PRNT_{80\%}$ titers >40 were protected against HTNV and DOBV (J.W. Hooper, unpublished data). Ultimately, it should be possible to develop a recombinant DNA vaccine containing a combination of immunogens that protect against all of the hantaviruses that cause human disease. Efforts to accomplish this goal are in progress.

3 Inactivated Hantavirus Vaccines in Asia

In recent years, there have been 50–150,000 cases of HFRS worldwide. More than 95% of these cases occur in Asia, mainly in China. To put these numbers in perspective, the total number of HPS cases reported since the first reported HPS outbreak in 1993 is only 555. The high number of cases of HFRS, especially in Asia, indicates an urgent need to develop a safe, effective vaccine against hantaviruses. This urgency has prompted researchers to produce vaccines by the same traditional inactivated-virus approaches that have been proven efficacious for other viral pathogens.

3.1 Rodent Brain-Derived Hantavirus Vaccines

Several groups throughout Asia have made inactivated hantavirus vaccines with rodent brain-derived virus (Yamanishi et al. 1988; Lee et al. 1990; Kim et al. 1991; Ahn et al. 1997). Most of the vaccines were made by methods similar to those used to prepare a Japanese encephalitis virus vaccine (Oya 1976). The virus is amplified to high titer in the brain tissue of suckling mice, hamsters, or rats. Viruses are purified by precipitation, ultra-filtration, and ultra-centrifugation. The infectious virus is inactivated by either formalin or β-propiolactone (BPL). Aluminum hydroxide is used as adjuvant in most of the vaccines.

In general the rodent brain-derived vaccines elicited good immunogenicity in rodent models as measured by IFAT and ELISA. Neutralizing antibody responses were also detected, but usually only after boosts. In the Republic of Korea, Lee et al. reported the development of vaccines derived from suckling rat brains infected with HTNV (strain ROK 84–105; Lee et al. 1990; Ahn et al. 1997). Infectious virus was inactivated by 0.05% formalin at, 4°C for 15 days. Each dose of the vaccine contained 0.5mg/ml of aluminum hydroxide as adjuvant. The immunogenicity of this vaccine was tested in mice. All of the vaccinated mice developed antibodies as measured by IFAT, but neutralizing antibody (low titer) was produced in only 40% of the vaccinated animals. Protective efficacy of the vaccine was tested by challenging the mice with homotypic virus and then measuring viral antigen in the lungs. All 30 vaccinated mice were protected from live virus challenge, whereas control mice were all infected (Lee et al. 1990). A human clinical trial demonstrated that 406 of 456 volunteers who received one dose of vaccine, and 333 of 336 who received two doses of vaccine, seroconverted as determined by IFAT; however, neutralizing antibody results were not reported (Lee et al. 1990).

A commercially produced version of the Korean rodent brain vaccine, Hantavax, contains HTNV (strain ROK 84–105), but is propagated in suckling mouse brains (rather than rat). In 64 human volunteers vaccinated with Hantavax subcutaneously (s.c.), 79% seroconverted as measured by IFAT and 62% by ELISA, 30 days after one vaccination (Cho and Howard 1999). One month after a second vaccination, the seroconversion rate increased to 97%. The conversion rates of neutralizing antibody were 13% and 75% 1month after the first vaccination and booster, respectively, with $PRNT_{50\%}$ titers ranging between 10 and 160. The vaccine is considered safe and only minor side effects were observed. Hantavax has been commercially available in South Korea since 1990.

An inactivated hantavirus vaccine was also developed in The People's Democratic Republic of Korea (Kim et al. 1991). The vaccine was derived from suckling rat or suckling hamster brains harvested from animals infected with a Hantaan-like virus (strain 821). The method of vaccine preparation was similar to that of the South Korean vaccines described above. Human volunteers received two doses of vaccine s.c. at 7- to 12-day intervals. Three weeks after the boost, seroconversion rates were 88.8% and 78.1% by IFAT and reversed passive hemagglutination inhibition assay (RPHI), respectively. No neutralizing antibody data were reported. The results of a field trial, reported to involve vaccination of 1.2 million people, indicated that the protective rates were between 88–100%.

3.2 Cell Culture-Derived Hantavirus Vaccines in China

Since the early 1980s, several independent research groups based in China have been working on the development of HFRS vaccines with inactivated hantaviruses grown in cell culture. Use of cell culture-derived virus avoids the use of large numbers of animals required to produce rodent brain-derived vaccines, and avoids other caveats such as the requirement to purify away myelin basic protein (which can cause allergic encephalomyelitis). A SEOV vaccine prepared in golden hamster kidney cells (GHKC), a HTNV vaccine prepared in Mongolian gerbil kidney cells (MGKC), and a bivalent (SEOV and HTNV) vaccine prepared in MGKC are among the hantavirus vaccines produced and licensed in China.

3.2.1 Monovalent Inactivated Seoul or Hantaan Virus Vaccines

GHKC grown in culture have been used to make viral vaccines in China for many years. Several vaccines such as Japanese encephalitis vaccine, rabies vaccine, and

Russian spring-summer encephalitis vaccine were derived from this cell line. A vaccine was developed from a Seoul-like virus (strain L99) that was originally isolated from the lung tissue of *Rattus losea* (Liu et al. 1984). This virus demonstrated wide cross-neutralization activity against several Chinese strains of hantavirus and grew to a high titer (10^7–$10^{7.5}$PFU/ml) in GHKC, which is similar to titers obtained in rodent brains.

Several methods of inactivating the GHKC-derived virus were evaluated including formalin, BPL, heating, and γ-irradiation (Liu et al. 1992). All of the methods completely inactivated the virus. The immunogenicity of the inactivated virus preparations was tested by injecting Wistar rats with the vaccines and then evaluating their antibody responses 4 weeks after vaccination. Tests to evaluate antibody responses included hemagglutination inhibition (HI), PRNT, anti-G2 and anti-N assays (Table 5). The virus preparation inactivated with γ-irradiation elicited the highest antibody response overall. The titers of neutralizing antibody induced by γ-irradiated virus were 2–4 times higher than those elicited by other methods. Thermal inactivation resulted in the lowest PRNT response but the highest anti-G2 antibody titer. After considering safety, expense, and production issues, formalin (0.025%v/v) was chosen as the vaccine inactivation method for the GHKC-derived vaccine. Similar studies were performed for other cell culture-derived vaccines, and either formalin or BPL were chosen as the method of inactivation. Incomplete inactivation of vaccine lots has not been reported.

Another group found that MGKC lines were susceptible to hantavirus infection. A Hantaan-like virus (strain Z10) was selected as the vaccine virus. BPL, at a final concentration of 0.05%, was used to inactivate the virus, and aluminum hydroxide (0.5mg/ml) was included as adjuvant.

The GHKC and MGKC vaccines were tested in rodents. Hamster challenge experiments were performed to test the protective efficacy of the GHKC-derived SEOV vaccine. Two doses of vaccine diluted to 1:25 protected more than 50% of hamsters from virus challenge (Hao et al. 1996; Table 6). Gerbil challenge experiments were performed to test the protective efficacy of the MGKC-derived HTNV

Table 5. Antibody responses in rats vaccinated[a] with hantavirus inactivated by different methods. (Modified from Liu et al. 1992)

Inactivation method[b]	Antibody titers														
	PRNT			HI			IFAT			Anti-G2[c]			Anti-N[c]		
Gamma ray (2×10^7rad, 12h)	640	640	640	160	320	160	5120	5120	640	8	64	32	1280	80	1280
Formalin (0.025%)	160	160	160	160	320	320	80	1280	2560	8	128	32	320	320	320
β-propiolactone (0.3%)	40	160	160	40	160		640	640			8	16	20	80	
Thermal (60°C, 60min)	<40	40	40	40	80	80	2560	2560	1280	64	64	128	20	320	320

[a] Wistar rats injected with vaccine three times at 2-week intervals;
[b] PRNT, plaque reduction neutralization test; HI, hemagglutination inhibition test; IFAT, indirect fluorescent antibody test;
[c] anti-G2 and anti-N responses were tested by ELISA.

Table 6. Protective effect of GHKC- and MGKC-derived vaccines in rodents. (Modified from Yu et al. 1990 and Hao et al. 1996)

Vaccine dilution	Protection rate[a](%)		
	GHKC-derived[b]		MGKC-derived[c]
	2 doses	3 doses	2 doses
1×	8/8 (100)	8/8 (100)	3/3 (100)
5×	7/8 (88)	7/8 (88)	2/3 (67)
25×	0/6 (0)	5/8 (63)	2/3 (67)
125×	0/6 (0)	2/8 (25)	0/3 (0)

[a] Protected/total tested, as measured by antigen in the lungs or spleens 7 days postchallenge as measured by IFAT;
[b] Hamsters received two or three doses of vaccine, i.p., at 1-week intervals and were then challenged with homotypic Seoul virus 3;
[c] Gerbils received two doses of vaccine, s.c., at a 2-week interval and were then challenged with homotypic Hantaan virus 2.

vaccine. Two doses of the vaccine diluted to 1:25 protected more than 50% of gerbils from virus challenge (Yu et al. 1990; Table 6).

To test the immunogenicity of the GHKC-derived SEOV vaccine in humans, phase I and II clinical trials were carried out. After three doses of vaccine, 97–100% of the volunteers had neutralizing antibody titers (Table 7) (Ren et al. 1996). To test the MGKC-derived HTNV vaccine, volunteers were vaccinated i.m. by different inoculation schedules, and their antibody responses were evaluated. After two inoculations at 1- or 4-weeks interval, only 40–50% vaccinees developed neutralizing antibody; however, 75–100% developed antibodies detectable by RPHI, IFAT, and ELISA. A three-dose schedule increased neutralizing antibody conversion to 92–100%, and antibodies detected by RPHI, IFAT, and ELISA were 83–100% (Zhu et al. 1991; Yu et al. 1992).

The GHKC-derived SEOV vaccine, MGKC-derived HTNV vaccine, and a suckling mouse brain (SMB)-derived HTNV vaccine were compared in a large human trial. Each vaccine was administered by different vaccination schedules according to manufacturer's instructions: GHKC-derived SEOV vaccine consisted of three vaccinations at 28- and 14-day intervals (primary vaccination) followed by a boost at 1 year; MGKC-derived HTNV vaccine consisted of three vaccinations at 7- and 14-day intervals (primary vaccination) followed by a boost at 1 year; and the SMB-derived HTNV vaccine consisted of three vaccinations at 2-week intervals (primary vaccination) followed by a boost at 1 year. The side effects in 55,699

Table 7. Human neutralizing antibody response to GHKC-derived vaccine. (From Ren et al. 1996)

Lot of vaccine[a]	Seroconversion rate[c] (%)	GMT[b]
92-1	33/33 (100)	94
92-2	37/38 (97.4)	96
92-3	36/36 (100)	100

[a] Vaccinees received three doses at 0, 28, 42 days intramuscularly;
[b] GMT, geometric mean titer;
[c] Seroconversion was determined by a Seoul virus plaque reduction neutralization test.

vaccinees who received three doses of vaccine i.m. were observed. The total incidence of side effects was 2.57%. SMB-derived vaccine produced the highest side effect rate of 7.29% (144/1962), GHKC-derived vaccine had a middle rate of 3.01% (687/22825), and MGKC-derived vaccine had the lowest rate of 1.90% (503/30912) (CHEN et al. 1999). All of the people in the trial regions were asked to visit a doctor for diagnosis when they had clinical symptoms suggesting HFRS. The diagnosis of HFRS in trial regions was made in local hospitals and confirmed by detection of IgM antibody in a central laboratory.

To evaluate the seroconversion rates of the vaccines, blood was drawn 2 weeks after they completed the primary vaccination schedule, 1 year after primary vaccination, and then 2 weeks, 1 year, 2 years, and 3 years after the 1-year boost. Two weeks after primary vaccination schedule, neutralizing antibody conversion rates were 51.09–82.46% (Table 8). One year later, the rates were 9.8–12.33%. Two weeks after the boost, rates were 61.90–80.13%. One year after the boost, the rates were 35.58–41.10%. Two years after the boost, the rates were 23.08–30.77%. Conversion rates of antibody detected by IFAT were higher than that of neutralizing antibody. These data revealed that hantavirus-specific antibody titers dropped significantly during the year after the primary vaccination. A boost 1 year after the primary vaccination resuscitated the antibody levels. After the 1-year boost, the drop in titer over the next 2 years was less precipitous than it was after the primary vaccination.

In the years after this large (> 100,000 participants) vaccination trial, the preventive efficacy of each vaccine was determined (CHEN et al. 1999). Four years

Table 8. Human seroconversion rates produced by three vaccines. (Modified from CHEN et al. 1999)

Vaccine	Test time[a]	Neutralizing assay (%)	IFAT (%)
GHKC[b]	2 weeks after primary vacc	82.46	65.73
	1 year after primary vacc	12.33	33.90
	2 weeks after booster vacc	80.13	85.00
	1 year after booster vacc	53.39	27.06
	2 years after booster vacc	37.04	41.49
	3 years after booster vacc	30.77	38.03
MGKC[c]	2 weeks after primary vacc	57.81	93.69
	1 year after primary vacc	9.87	17.76
	2 weeks after booster vacc	70.71	89.04
	1 year after booster vacc	35.58	50.00
	2 years after booster vacc	21.77	32.88
	3 years after booster vacc	23.08	39.53
SMB[d]	2 weeks after primary vacc	51.09	84.27
	1 year after primary vacc	10.48	18.06
	2 weeks after booster vacc	61.90	83.50
	1 year after booster vacc	41.10	41.77
	2 years after booster vacc	25.00	12.68

[a] Vacc, vaccination;
[b] GHKC-derived Seoul virus vaccine, administered on day 0, 14, 28 (primary vaccination), boost at 1 year;
[c] MGKC-derived Hantaan virus vaccine, administered on day 0, 7, and 14 (primary vaccination), boost at 1 year;
[d] SMB-derived Hantaan virus vaccine, administered on day 0, 14, 28 (primary vaccination), boost at 1 year.

Table 9. Preventive rates of three inactivated hantavirus vaccines. (Modified from CHEN et al. 1999)

Vaccine	Time after vaccination	Incidence per 100,000 (cases/total)[a]		Preventive rate (%)
		Vaccinated	Unvaccinated	
GHKC[b]	1st year	2.45 (1/40757)	141.61 (67/47313)	98
	2nd year	2.48 (1/40289)	110.17 (52/47201)	98
	3rd year	2.48 (1/40289)	97.46 (46/47201)	97
	4th year	2.48 (1/40289)	63.56 (30/47201)	96
MGKC[c]	1st year	6.06 (3/44104)	153.07 (96/62718)	96
	2nd year	4.62 (2/43285)	55.82 (35/62702)	92
	3rd year	2.36 (1/42304)	38.28 (24/62702)	94
	4th year	2.36 (1/42304)	14.35 (9/67202)	84
SMB[d]	1st year	7.55 (2/26492)	115.28 (40/34699)	93
	2nd year	0.00 (0/24111)	49.14 (15/30525)	100
	3rd year	4.85 (1/20626)	45.86 (14/30525)	89
	4th year	0.00 (0/20626)	9.83 (3/30525)	100

[a] number of HRFS cases/number of participants in trial;
[b] GHKC-derived Seoul virus vaccine, administered on day 0, 14, 28 (primary vacc), boost at 1 year;
[c] MGKC-derived Hantaan virus vaccine, administered on day 0, 7, and 14 (primary vacc), boost at 1 year;
[d] SMB-derived Hantaan virus vaccine, administered on day 0, 14, 28 (primary vacc), boost at 1 year.

after the primary vaccination, average preventive rates were >90% (Table 9). In the trial, 14 cases of HFRS were confirmed among all of vaccinees, and 431 cases of HFRS among the control group. There were no significant differences of clinical manifestations between patients from vaccination and control groups. So far, there is no evidence of antibody-dependent enhancement of infection or disease.

3.2.2 Bivalent Inactivated Seoul and Hantaan Virus Vaccines

To avoid the need to use two vaccines in China, one to Seoul- and one to Hantaan-like viruses, bivalent vaccines were developed. One bivalent vaccine consists of a GHKC-derived Hantaan-like (strain 84Fli), and Seoul-like virus (strain L99) (D.X. Li and C.S. Hang, unpublished data). Another bivalent vaccine consists of MGKC-derived Seoul-like virus, strain Z37, and Hantaan-like virus, strain Z10, inactivated with BPL (DONG et al. 1998). The protocols for preparing both bivalent vaccines are similar to that of monovalent vaccine production. Both bivalent vaccines have finished phase II clinical trials.

Each phase II trial included 1,000 volunteers that were divided into two groups located in southern and northern China. The volunteers received three doses of 1ml of vaccine i.m. at 2-week intervals. A subset of volunteers, 200 receiving the GHKC-derived vaccine and 260 receiving the MGKC-derived vaccine, was monitored closely for side effects. Side effect incidences were 2% and 4% for the GHKC- and MGKC-derived vaccine, respectively. Two weeks after the final vaccination, the vaccinees' blood was drawn and evaluated for HTNV and SEOV neutralizing antibody by PRNT. The seroconversion rates (titers >1:10) to HTNV and SEOV vaccinated with GHKC-derived vaccine were 89% and 92%, and with MGKC-derived vaccine were 87.6% and 96.3%, respectively.

To date, five manufacturers in China have been approved to produce commercial vaccines against hantaviruses: two producing GHKC-derived vaccines, two producing MGKC-derived vaccines, and one producing SMB-derived vaccine. Together, 4 million doses of hantavirus vaccine are produced annually in China. Studies thus far indicate that it should be possible to manufacture a bivalent vaccine that protects against both types of HFRS that are prevalent in China.

4 Concluding Remarks

HFRS was essentially unknown to Western medicine until the Korean War, when over 2,400 United Nations soldiers contracted the disease (Smadel 1953). The wartime epidemics from 1951 to 1954 began a research effort aimed at treating and preventing HFRS that has continued into the 21st century. During this time, hantaviruses were identified as not only the causative agent of HFRS in Asia, but also of nephropathica epidemica in Eurasia, and of HPS in the Americas. Education and rodent-control measures have had a significant impact on the prevalence of disease; nevertheless, vaccines are needed to protect populations with a high risk of exposure.

Because hantaviruses were not known to cause disease in the Americas until very recently, the majority of hantavirus vaccine research in the United States was performed under the auspices of the US military with the goal of protecting personnel deployed in hantavirus-endemic regions. The recent emergence of HPS-associated hantaviruses in the Americas, especially a strain that apparently can be transmitted from person-to-person (Wells et al. 1997), has provided additional impetus to develop multivalent hantavirus vaccines. In the United States, killed hantavirus vaccine strategies that involve the growth, purification, and inactivation of large quantities of biosafety level-3 viruses have been avoided. Rather, the vaccine efforts have sought to elicit protective immune responses by using recombinant DNA approaches that involve vaccination with specific hantavirus proteins. Several of these approaches (e.g., recombinant vaccinia viruses and naked DNA) allow expression of hantavirus proteins within the cells of vaccinees. These approaches are likely to elicit a qualitatively and quantitatively different immune response than inactivated virus delivered with adjuvant. For example, naked DNA gene gun vaccination of hamsters with the SEOV M gene elicited neutralizing antibody titers several fold higher than those elicited by inactivated virus vaccines (Table 3). Future efforts will investigate whether naked DNA vaccination with other hantavirus M genes can elicit high-titer neutralizing antibodies, and whether neutralizing antibody responses can be elicited in non-human primates and humans.

Vaccine development efforts in Asia, where HFRS is a major health problem, have primarily focused on inactivated virus vaccines. These efforts have resulted in the production of several vaccines that have progressed into large-scale human trials and licensure. The vaccines appear to be effective against the two most important hantaviruses in Asia, HTNV and SEOV, with minor side effects. Confirmation of the vaccines' efficacy will become more apparent as greater proportions of the population are vaccinated. Optimization of vaccination schedules and advances in adjuvants may further increase the efficacy of inactivated hantavirus vaccines.

References

Ahn CN (1997) Hemorrhagic fever with renal syndrome (HFRS) vaccine in Korea WHO working group on prevention and control of Hantavirus infections. (Abstract) Seoul 5–6 Nov 1997

Arikawa J, Yao JS, Yoshimatsu K, Takashima I, Hashimoto N (1992) Protective role of antigenic sites on the envelope protein of Hantaan virus defined by monoclonal antibodies. Arch Virol 126, 271–281

Bayard V, et al. (2000) Hantavirus pulmonary syndrome: Panama, 1999–2000. MMWR 49:205–207

Bharadwaj M, Lyons RC, Wortman IA, Hjelle B (1999) Intramuscular inoculation of Sin Nombre hantavirus cDNAs induces cellular and humoral immune responses in BALB/c mice. Vaccine 17:2836–2843

Chen HX, Luo ZZ, Zhang JJ (1999) Study on epidemiological efficacy and immunologic strategy of vaccines against hemorrhagic fever with renal syndrome Chinese Public Health 15:561–8 (in Chinese)

Cho HW, Howard CR (1999) Antibody responses in humans to an inactivated hantavirus vaccine (Hantavax) Vaccine 4:2569–75

Chu YK, Jennings GB, Schmaljohn CS (1995) A vaccinia virus-vectored Hantaan virus vaccine protects hamsters from challenge with Hantaan and Seoul viruses but not Puumala virus. J Virol 69:6417–6423

Dong GM, An Q, Zhu ZY (1998) Expanded human trail of the bivalent HFRS inactivated vaccine made from *M. gerbil* kidney tissue culture. Chin J Microbiol Immunol 18:453–456

Dubensky T, Driver D, Polo J, Belli B, Latham E, Ibanex C, Chada S, Brumm D, Banks T, Mento S, Jolly D, Chang S (1996) Sindbis virus DNA-based expression vectors: utility for in vitro and in vivo gene therapy. J Virol 70:508–519

Frolov I, Hoffman TA, Pragai BM, Dryga SA, Huang HV, Schlesginer S, Rice CM (1996) Alphavirus-based expression vectors: strategies and applications. Proc Natl Sci USA 93:11 371–11 377

Hao FY, Hui LG, Zhao XL, Wang XH, Han L, Huang YC (1996) Efficacy test for inactivated epidemic hemorrhage fever vaccine using golden hamsters. Chin J Biol 9:69–72 (in Chinese)

Hooper JW, Kamrud KI, Elgh F, Custer D, Schmaljohn CS (1999) DNA vaccination with hantavirus M segment elicits neutralizing antibodies and protects against Seoul virus infection. Virology 255: 269–278

Kamrud KI, Hooper JW, Elgh F, Schmaljohn CS (1999) Comparison of protective efficacy of naked DNA, DNA-based replicon, and packaged Sindbis replicon vectors expressing hantavirus structural genes in hamsters. Virology 263:209–219

Kim RJ, Ru C, Kim GM (1991) The special prevention of HFRS in P.D.R of Korea. Chinese Clin Exper Virol 4:487–92 (in Chinese)

Lee H, Ahn C, Song J, Baek L, Seo T, Park S (1990) Field trial of an inactivated vaccine against hemorrhagic fever with renal syndrome in humans. Arch Virol (Suppl 1):35–47

Liu PQ, Liao HX, Fu JL (1984) Isolation of epidemic hemorrhagic fever viruses from Rattus losea and R. Confucianus, and their antigenic identification. Bull Jiangxi Med Coll 3:1(in Chinese)

Liu WM, Song G, Zhang QF (1992) Comparative studies on inactivating methods for production of inactivated cell culture vaccine against epidemic hemorrhagic fever. Chin J Virol 3:141–146 (in Chinese)

Lundkvist Å, Kallio-Kokko H, Brus-Sjölander K, Lankinen H, Niklasson B, Vaheri A, Vapalahti O (1996) Characterization of Puumala virus nucleocapsid protein: identification of B-cell epitopes and domains involved in protective immunity. Virology 216:397–406

McClain DJ, Summers PL, Harrison SA, Schmaljohn AL, Schmaljohn CS (2000) Clinical evaluation of a vaccinia-vectored Hantaan virus vaccine. J Med Virol 60:77–85

Oya A (1976) Japanese encephalitis vaccine: the vaccination. Int Medical Found Japan 1976:69–72

Pensiero MN, Jennings GB, Schmaljohn CS, Hay J (1988) Expression of the Hantaan virus M genome segment by using a vaccinia virus recombinant. J Virol 62:696–702

Ren K, Lu QX, Song G, Huang YC, Sun RH, Wang S (1996) Serological efficacy of the inactivated golden hamster kidney cell vaccine against hemorrhagic fever with renal syndrome in human trial. Chin J Exp Clin Virol 10:10–15 (in Chinese)

Robinson HL, Torres CA (1997) DNA vaccines. Semin Immunol 9:271–283

Schmaljohn CS (1996) Molecular Biology of Hantaviruses. In: Elliott RM (ed) The *Bunyaviridae*. New York, Plenum Press pp 63–90

Schmaljohn CS, Chu YK, Schmaljohn AL, Dalrymple JM (1990) Antigenic subunits of Hantaan virus expressed by baculovirus and vaccinia virus recombinants. J Virol 64:3162–3170

Schmaljohn CS, Hasty SE, Dalrymple JM (1992) Preparation of candidate vaccinia-vectored vaccines for haemorrhagic fever with renal syndrome. Vaccine 10:10–13

Smadel JE (1953) Epidemic hemorrhagic fever. Am J Pub Hlth 43:13327–13335

Ulrich R, Lundkvist Å, Meisel H, Koletzki D, Brus Sjölander K, Gelderblom HR, Borisova G, Schnitzler P, Darai G, Kruger DH (1998) Chimaeric HBV core particles carrying a defined segment of Puumala hantavirus nucleocapsid protein evoke protective immunity in an animal model. Vaccine 16:272–280

Ulrich R, Koletzki H, Lachmann S, Lundkvist A, Zankl A, Kazaks A, Kurth A, Gelderblom R, Borisova G, Meisel H, Kruger DH (1999) New chimaeric hepatitis B virus core particles carrying hantavirus (serotype Puumala) epitopes: immunogenicity and protection against virus challenge. J Biotechnol 73:141–153

Wells RM, Sosa Estani S, Yadon ZE, Enria D, Padula P, Pini N, Mills JN, Peters CJ, Segura EL (1997) An unusual hantavirus outbreak in southern Argentina: person-to-person transmission? Hantavirus Pulmonary Syndrome Study Group for Patagonia. Emerg Infect Dis Apr-Jun3:171–174

Xu X, Ruo SL, McCormick JB, Fisher-Hoch SP (1992) Immunity to Hantavirus challenge in Meriones unguiculatus induced by vaccinia-vectored viral proteins. Am J Trop Med Hyg 47:397–404

Yamanishi K, Tanishita O, Tamura M, Asada H, Kondo K, Takagi M, Yoshida I, Konobe T, Fukai K (1988) Development of inactivated vaccine against virus causing haemorrhagic fever with renal syndrome. Vaccine 3:278–282

Yoshimatsu K, Yoo YC, Yoshida R, Ishihara C, Azuma I, Arikawa J (1993) Protective immunity of Hantaan virus nucleocapsid and envelope protein studied using baculovirus expressed proteins. Arch Virol 130:365–376

Yu YX, Yao XJ, Dong GM, An Q, Li ZP (1990) Antibody response of inactivated HFRS vaccine to homologous and heterologous types of the virus. Chin J Biol 3:14–16

Yu YX, Liu WX, Nei ZL (1992) Neutralizing antibody response in humans immunized with Meriones gerbil kidney tissue culture inactivated HFRS vaccine assesed by two methods. Virol Sin 7:176–180

Zhu ZY, Zeng RF, Yu XY (1991) Efficacy of inactivated EHF vaccine in clinical trial. Virol Sin 6: 315–319

Subject Index

Current Topics in Microbiology and Immunology

Volumes published since 1989 (and still available)

Vol. 213/III: **Günthert, Ursula; Schlag, Peter M.; Birchmeier, Walter (Eds.):** Attempts to Understand Metastasis Formation III. 1996. 14 figs. XV, 262 pp. ISBN 3-540-60682-3

Vol. 214: **Kräusslich, Hans-Georg (Ed.):** Morphogenesis and Maturation of Retroviruses. 1996. 34 figs. XI, 344 pp. ISBN 3-540-60928-8

Vol. 215: **Shinnick, Thomas M. (Ed.):** Tuberculosis. 1996. 46 figs. XI, 307 pp. ISBN 3-540-60985-7

Vol. 216: **Rietschel, Ernst Th.; Wagner, Hermann (Eds.):** Pathology of Septic Shock. 1996. 34 figs. X, 321 pp. ISBN 3-540-61026-X

Vol. 217: **Jessberger, Rolf; Lieber, Michael R. (Eds.):** Molecular Analysis of DNA Rearrangements in the Immune System. 1996. 43 figs. IX, 224 pp. ISBN 3-540-61037-5

Vol. 218: **Berns, Kenneth I.; Giraud, Catherine (Eds.):** Adeno-Associated Virus (AAV) Vectors in Gene Therapy. 1996. 38 figs. IX,173 pp. ISBN 3-540-61076-6

Vol. 219: **Gross, Uwe (Ed.):** Toxoplasma gondii. 1996. 31 figs. XI, 274 pp. ISBN 3-540-61300-5

Vol. 220: **Rauscher, Frank J. III; Vogt, Peter K. (Eds.):** Chromosomal Translocations and Oncogenic Transcription Factors. 1997. 28 figs. XI, 166 pp. ISBN 3-540-61402-8

Vol. 221: **Kastan, Michael B. (Ed.):** Genetic Instability and Tumorigenesis. 1997. 12 figs.VII, 180 pp. ISBN 3-540-61518-0

Vol. 222: **Olding, Lars B. (Ed.):** Reproductive Immunology. 1997. 17 figs. XII, 219 pp. ISBN 3-540-61888-0

Vol. 223: **Tracy, S.; Chapman, N. M.; Mahy, B. W. J. (Eds.):** The Coxsackie B Viruses. 1997. 37 figs. VIII, 336 pp. ISBN 3-540-62390-6

Vol. 224: **Potter, Michael; Melchers, Fritz (Eds.):** C-Myc in B-Cell Neoplasia. 1997. 94 figs. XII, 291 pp. ISBN 3-540-62892-4

Vol. 225: **Vogt, Peter K.; Mahan, Michael J. (Eds.):** Bacterial Infection: Close Encounters at the Host Pathogen Interface. 1998. 15 figs. IX, 169 pp. ISBN 3-540-63260-3

Vol. 226: **Koprowski, Hilary; Weiner, David B. (Eds.):** DNA Vaccination/Genetic Vaccination. 1998. 31 figs. XVIII, 198 pp. ISBN 3-540-63392-8

Vol. 227: **Vogt, Peter K.; Reed, Steven I. (Eds.):** Cyclin Dependent Kinase (CDK) Inhibitors. 1998. 15 figs. XII, 169 pp. ISBN 3-540-63429-0

Vol. 228: **Pawson, Anthony I. (Ed.):** Protein Modules in Signal Transduction. 1998. 42 figs. IX, 368 pp. ISBN 3-540-63396-0

Vol. 229: **Kelsoe, Garnett; Flajnik, Martin (Eds.):** Somatic Diversification of Immune Responses. 1998. 38 figs. IX, 221 pp. ISBN 3-540-63608-0

Vol. 230: **Kärre, Klas; Colonna, Marco (Eds.):** Specificity, Function, and Development of NK Cells. 1998. 22 figs. IX, 248 pp. ISBN 3-540-63941-1

Vol. 231: **Holzmann, Bernhard; Wagner, Hermann (Eds.):** Leukocyte Integrins in the Immune System and Malignant Disease. 1998. 40 figs. XIII, 189 pp. ISBN 3-540-63609-9

Vol. 232: **Whitton, J. Lindsay (Ed.):** Antigen Presentation. 1998. 11 figs. IX, 244 pp. ISBN 3-540-63813-X

Vol. 233/I: **Tyler, Kenneth L.; Oldstone, Michael B. A. (Eds.):** Reoviruses I. 1998. 29 figs. XVIII, 223 pp. ISBN 3-540-63946-2

Vol. 233/II: **Tyler, Kenneth L.; Oldstone, Michael B. A. (Eds.):** Reoviruses II. 1998. 45 figs. XVI, 187 pp. ISBN 3-540-63947-0

Vol. 234: **Frankel, Arthur E. (Ed.):** Clinical Applications of Immunotoxins. 1999. 16 figs. IX, 122 pp. ISBN 3-540-64097-5

Vol. 235: **Klenk, Hans-Dieter (Ed.):** Marburg and Ebola Viruses. 1999. 34 figs. XI, 225 pp. ISBN 3-540-64729-5

Vol. 236: **Kraehenbuhl, Jean-Pierre; Neutra, Marian R. (Eds.):** Defense of Mucosal Surfaces: Pathogenesis, Immunity and Vaccines. 1999. 30 figs. IX, 296 pp. ISBN 3-540-64730-9

Vol. 237: **Claesson-Welsh, Lena (Ed.):** Vascular Growth Factors and Angiogenesis. 1999. 36 figs. X, 189 pp. ISBN 3-540-64731-7

Vol. 238: **Coffman, Robert L.; Romagnani, Sergio (Eds.):** Redirection of Th1 and Th2 Responses. 1999. 6 figs. IX, 148 pp. ISBN 3-540-65048-2

Vol. 239: **Vogt, Peter K.; Jackson, Andrew O. (Eds.):** Satellites and Defective Viral RNAs. 1999. 39 figs. XVI, 179 pp. ISBN 3-540-65049-0

Vol. 240: **Hammond, John; McGarvey, Peter; Yusibov, Vidadi (Eds.):** Plant Biotechnology. 1999. 12 figs. XII, 196 pp. ISBN 3-540-65104-7

Vol. 241: **Westblom, Tore U.; Czinn, Steven J.; Nedrud, John G. (Eds.):** Gastroduodenal Disease and Helicobacter pylori. 1999. 35 figs. XI, 313 pp. ISBN 3-540-65084-9

Vol. 242: **Hagedorn, Curt H.; Rice, Charles M. (Eds.):** The Hepatitis C Viruses. 2000. 47 figs. IX, 379 pp. ISBN 3-540-65358-9

Vol. 243: **Famulok, Michael; Winnacker, Ernst-L.; Wong, Chi-Huey (Eds.):** Combinatorial Chemistry in Biology. 1999. 48 figs. IX, 189 pp. ISBN 3-540-65704-5

Vol. 244: **Daëron, Marc; Vivier, Eric (Eds.):** Immunoreceptor Tyrosine-Based Inhibition Motifs. 1999. 20 figs. VIII, 179 pp. ISBN 3-540-65789-4

Vol. 245/I: **Justement, Louis B.; Siminovitch, Katherine A. (Eds.):** Signal Transduction and the Coordination of B Lymphocyte Development and Function I. 2000. 22 figs. XVI, 274 pp. ISBN 3-540-66002-X

Vol. 245/II: **Justement, Louis B.; Siminovitch, Katherine A. (Eds.):** Signal Transduction on the Coordination of B Lymphocyte Development and Function II. 2000. 13 figs. XV, 172 pp. ISBN 3-540-66003-8

Vol. 246: **Melchers, Fritz; Potter, Michael (Eds.):** Mechanisms of B Cell Neoplasia 1998. 1999. 111 figs. XXIX, 415 pp. ISBN 3-540-65759-2

Vol. 247: **Wagner, Hermann (Ed.):** Immunobiology of Bacterial CpG-DNA. 2000. 34 figs. IX, 246 pp. ISBN 3-540-66400-9

Vol. 248: **du Pasquier, Louis; Litman, Gary W. (Eds.):** Origin and Evolution of the Vertebrate Immune System. 2000. 81 figs. IX, 324 pp. ISBN 3-540-66414-9

Vol. 249: **Jones, Peter A.; Vogt, Peter K. (Eds.):** DNA Methylation and Cancer. 2000. 16 figs. IX, 169 pp. ISBN 3-540-66608-7

Vol. 250: **Aktories, Klaus; Wilkins, Tracy, D. (Eds.):** Clostridium difficile. 2000. 20 figs. IX, 143 pp. ISBN 3-540-67291-5

Vol. 251: **Melchers, Fritz (Ed.):** Lymphoid Organogenesis. 2000. 62 figs. XII, 215 pp. ISBN 3-540-67569-8

Vol. 252: **Potter, Michael; Melchers, Fritz (Eds.):** B1 Lymphocytes in B Cell Neoplasia. 2000. XIII, 326 pp. ISBN 3-540-67567-1

Vol. 253: **Gosztonyi, Georg (Ed.):** The Mechanisms of Neuronal Damage in Virus Infections of the Nervous System. 2001. approx. XVI, 270 pp. ISBN 3-540-67617-1

Vol. 254: **Privalsky, Martin L. (Ed.):** Transcriptional Corepressors. 2001. 25 figs. XIV, 190 pp. ISBN 3-540-67569-8

Vol. 256: **Hirai, Kanji (Ed.):** Marek's Disease. 2001. 22 figs. XII, 294 pp. ISBN 3-540-67798-4

Lightning Source UK Ltd.
Milton Keynes UK
UKHW020739091218
333681UK00012B/441/P

9 783540 410454